Jim BABCOCK

The V Series
Recommendations

McGraw-Hill Series-Computer Communications

The V Series Recommendations

Standards for Data Communications over the Telephone Network

Second Edition

Uyless Black

McGraw-Hill, Inc.

New York San Francisco Washington, D.C. Auckland Bogotá
Caracas Lisbon London Madrid Mexico City Milan
Montreal New Delhi San Juan Singapore
Sydney Tokyo Toronto

Product or brand names used in this book may be trade names or trademarks. Where we believe that there may be proprietary claims to such trade names or trademarks, the name has been used with an initial capital or it has been capitalized in the styled used by the name claimant. Regardless of the capitalization used, all such names have been used in an editorial manner without any intent to convey endorsement of or other affiliation with the name claimant. Neither the author nor the publisher intends to express any judgment as to the validity or legal status of any such proprietary claims.

Library of Congress Cataloging-in-Publication Data

Black, Uyless D.
    The V series recommendations : standards for data communications
over the telephone network / by Uyless Black—2nd ed.
        p.   cm.—(Uyless Black series on computer communications)
    Includes index.
    ISBN 0-07-005592-0
    1. Computer network protocols—Standards.   2. Telephone systems-
-Standards.   I. Title.   II. Series.
    TK5105.5.B5673   1995
    004.6'2—dc20   94-39989
                                                                              CIP

1 2 3 4 5 6 7 8 9 0 DOC/DOC   9 8 7 6 5

ISBN 0-07-005592-0

The sponsoring editor was Brad Schepp, the manuscript editor was John C. Baker, and the executive editor was Robert Ostrander. The production supervisor was Katherine Brown. This book was set in ITC Century Light. It was composed in Blue Ridge Summit, Pa.

Printed and bound by R. R. Donnelley & Sons Company, Crawfordsville, Indiana.

MH95
0055920

To my friends Doug and Nita Hammock

Contents

Acknowledgments

A special note of thanks is due to the CCITT for their ongoing efforts in promoting and publishing standards for data communications systems and networks. These standards can only help companies and organizations in their efforts to increase productivity, reduce the complexity of their communications interfaces, and reduce overhead.

This book has been prepared independently of CCITT and reflects the views of the author and not necessarily those of CCITT.

Please be aware that the ITU/CCITT are the copyright holders for the V Series Recommendations. The excerpting and reproduction of any material is authorized by the ITU/CCITT organizations, the copyright holders. The choice of any excerpts is that of the author and does not affect ITU/CCITT in any way.

The full text of these recommendations can be obtained from the ITU Sales Section, Place des Nations, Geneva, Switzerland. Please refer to Appendix C for more information.

Preface

The idea for this book came from a number of my clients who expressed a need for a book that provided a tutorial on the V Series Recommendations as well as an abbreviated reference guide on the subject. It seemed a good idea to me also. Therefore, I began the preparation of the manuscript in conjunction with a series of lectures that I made on the subject in Europe. After refining the material with seminar delegates and clients, I submitted it to my publisher for inclusion in this series.

The book contains information on each of the V Series Recommendations. Although many of these Recommendations might not pertain to a person's job or might not be of interest, practically any data communications professional must come to grips with many of these standards and, therefore, the amount of text in the Recommendations themselves still is somewhat overwhelming. I have attempted to abbreviate the Recommendations but still provide enough information on each so that the explanation is not superficial. In general, I also have avoided engineering details that are best left to the source documents.

It must be understood that this book is meant as a tutorial and a general guide on the subject. You cannot design or implement a system by only reading this book. It should not be used as a replacement for your reading and studying the actual ITU-T V Series Recommendation; there is no substitute for the actual source specification. Notwithstanding, it is hoped that this material will make the V Series documents more understandable to the reader.

The major changes from the first edition of the book reflect the revisions and new Recommendations made to the ITU-T Blue Book (actually Blue Books) that were published in 1988.

Notes for the Reader

The International Telegraph and Telephone Consultative Committee (CCITT) has been reorganized and renamed (appendix 1A in chapter 1 provides more details). It now is called the International Telecommunication Union-Telecommunication Standardization Sector (or ITU-T).

I have delayed preparing the second edition of this book in anticipation of the V.34 Recommendation being approved early in 1994, but it has been postponed several times. I decided to proceed with this edition and have included a description of the work that has been completed thus far on V.34.

Periodically, the ITU-T declares outdated Recommendations "no longer in force" (NLIF), which means that they are not supported by the ITU-T, and the ITU-T recommends that they be discontinued in vendor products. Notwithstanding, they still might be used in some vendor products and might be important to the reader. Therefore, this book includes the more recent NLIF Recommendations. In the chapter in which they are described, they are identified with the initials NLIF for: no longer in force.

The V Series are undergoing revisions, deletions, and additions continuously. For those Recommendations that have not been completed (as of this writing), they will be noted in the appropriate chapter with UCFA.

The initials of NLIF and UCFA are created for this book and are not used by the ITU-T.

The term *bit/s* is used to describe the transmission rate in bits per second. Due to space constraints, in some figures and tables, the term *b/s* also is used to describe the same operations.

In some of the figures, I have used abbreviated titles or have shortened the titles of the various V Series Recommendations. Appendix C provides a list of the full titles and ordering information.

Introduction to the V Series

This book examines the International Telecommunication Union-Telecommunication Standardization Sector (ITU-T) Recommendations pertaining to the V Series (see Appendix 1A for a discussion of the ITU-T). The focus of the material is the "Blue Book" (actually Blue Books) and subsequent updates to it.

The V Series Recommendations (also called *standards* by many people) have become some of the most widely used specifications in the world for defining how data are exchanged between computers and communications equipment such as modems and multiplexers. In the past, the V Series Recommendations were used principally in Europe because of the influence of the Postal, Telephone, and Telegraph (PTT) administrations in each European country. However, with the growing recognition of the need for international communications standards, the V Series Recommendations have found their way into most countries of the world and into practically all vendors' products. Their use has paved the way for easier, more efficient, and less costly communications between users' computers, terminals, and other data processing machines.

The V Series Recommendations are covered in over 500 pages, and the volume and complexity of the V Series material present problems in staying abreast of these important international standards. Until recently, the ITU-T revised the V Series Recommendations every four years. (You will see later that the ITU-T has altered this approach to keep the standards more current.) Unless you have a considerable amount of time to devote to detailed research, it is impossible to read all of these standards.

It is my intention that this book will aid you in learning about these im-

portant standards. This book provides you with a tutorial on each of the recommended standards. It also is structured to provide a reference guide to
each of the V Series Recommendations.

Purpose of the V Series

The V Series Recommendations are titled *Data Communication Over the
Telephone Network*. The title describes their functions: to define conventions and procedures for the transfer of data using the public telephone network. Because the vast majority of telephone systems use analog signalling
to support voice communications, a substantial part of the V Series
Recommendations is devoted to defining the conventions for converting
digital data signals (binary digits, or bits) into analog signals, and vice versa.

Relationship of the V Series to a Communications Link

The V Series Recommendations include the descriptions of the physical interfaces between the communicating machines. It sometimes is assumed
that a physical interface encompasses only the interchange circuits between the data terminal equipment (DTE; the user device) and the data circuit-terminating equipment (DCE; the communications device). While this
view is correct for some products and standards, the physical level also includes the signalling between the two DCEs (see Figure 1.1).

Most of the V Series physical-level protocols include information on the
DTE-to-DCE interface, the DCE-to-DCE interface, or both sides of the inter-
face. Initially this approach might be somewhat confusing, but the ITU-T V
Series protocols are rather explicit and quite clear as to which side of the

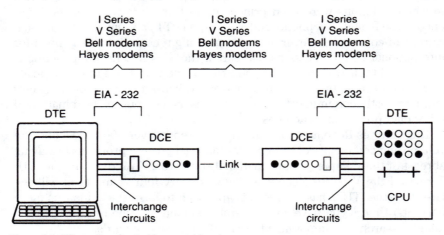

Figure 1.1 Where the physical layer resides.

DCE pertains to each recommendation. As you shall see in later chapters, the widely used V.28 and V.24 Recommendations are pertinent to the DTE-to-DCE side of the physical-level interface. On the other hand, the V Series voice-band and wide-band modems pertain to the DCE-to-DCE side of the interface. The confusion arises because the voice-band and wide-band modem specifications *also* stipulate the use of other V Series Recommendations such as V.24 and V.28, thus defining both sides of the interface.

Also, you will see that each of the modem specifications that are contained in the V Series Recommendations define both sides of the interface. As a consequence, each standard is a "complete package" to be read without the need to study the other recommendations unless you need more detailed information on the specifics of the interfaces that deal with the specifications such as V.28, V.24, or some other recommendations.

As shown in Figure 1.1, other prominent standards, such as EIA-232-E, encompass only the DTE-to-DCE side of the interface. Even though EIA-232-E does not specify the DCE-to-DCE exchange, many vendors use the relevant portion of the ITU-T V Series Recommendations and/or their own specifications to describe this part of the physical level interface. For example, this approach has been taken for the Hayes modem products.

The I Series Recommendations specify a digital physical-level interface. You can refer to appendix B for a brief tutorial on the integrated services digital network (ISDN).

Clarification of the term DCE

While the terms DTE and DCE are useful to depict a conceptual operation, in many actual situations, the functions of the DTE and DCE are located in one machine. For example, a mainframe computer or front-end processor might contain certain functions of DTEs and DCEs on the same line card. As another example, portable personal computers (such as laptops) place the modem inside the case of the personal computer. You also should be aware that the term DCE is used by many manufacturers to denote a network interface with the DTE, even though the network exchange (such as a packet switch or a circuit switch) might reside in a different location and in a separate piece of equipment.

Structure of the V Series

The ITU-T Study Group 14 is responsible for the V Series. Table 1.1 lists the broad categories of functions and services that are described in the V Series. Figure 1.2 provides an illustration of the organization of the V Series. The recommendations are organized into six sections. Each of these sections is further divided into the specific recommendation. This approach facilitates the organization and use of the documents.

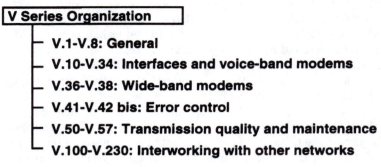

Figure 1.2 The V Series organization.

TABLE 1.1 Organization of the V Series

Section number and name	Description
1. General	General overview descriptions of coding, symbols, signalling rates, and power levels that are used over the telephone network
2. Interfaces and voice-band modems	Detailed descriptions of the interfaces between modems and their DTEs, as well as the signalling conventions between the modems over voice-band frequencies
3. Wide-band modems	Detailed descriptions of the interfaces between wide-band modems and their DTEs, as well as the signalling conventions between the modems
4. Error control	Descriptions of conventions to obtain error-detection and error-correcting services
5. Transmission quality	Specifications on testing methods, noise and measurement, and maintenance limits on telephone circuits
6. Interworking with other networks	Specifications for interworking telephone networks with packet and ISDN networks

The V Series Recommendations are identified by titles (which sometimes are rather lengthy) and a two- or three-digit number, which is preceded by the letter *V*. The full titles of each recommendation are provided in the appropriate chapter and in appendix C. For the present discussion, the following V numbers are associated with the six categories:

- General: V.1 through V.8
- Interfaces and voice-band modems: V.10 through V.34

- Wide-band modems: V.35 through V.38
- Error control: V.40 through V.42
- Transmission quality and maintenance: V.50 through V.57
- Interworking with other networks: V.100 through V.230

For purposes of organization and simplicity, I have divided the ITU-T section titled "Interfaces and voice-band modems" into two chapters in this book.

Commonly Used Terms in the V Series Recommendations

This section introduces and explains some terms that are used in the V Series Recommendations. They are rather basic, but the ITU-T documents assume the reader understands them.

Figure 1.3 depicts the relationship of the user access line to the networks. In this figure, user A connects its DTE, which is an end user computer, to a network through the *user access line* and the switch. ITU-T uses the term *data circuit-terminating equipment* (DCE) in two contexts. In this figure, it is used to illustrate the user-network interface exit machine for a packet network. In later discussions, it is used to illustrate a device, such as a modem. This figure also illustrates the internetworking unit (IWU), which is used to interconnect networks.

The network on the left of the figure is a packet data network. It uses conventional switching technology to route traffic through the network (the packet switches). The network on the right is an example of a broad-

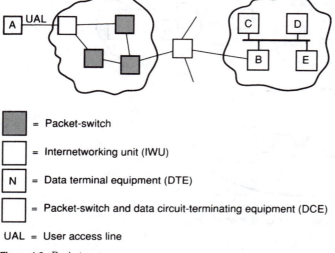

      = Packet-switch

      = Internetworking unit (IWU)

N   = Data terminal equipment (DTE)

      = Packet-switch and data circuit-terminating equipment (DCE)

UAL = User access line

Figure 1.3 Basic terms.

cast network (in this example, a local area network, or LAN). The traffic is sent to all of the devices that are attached to the network (DTEs B, C, D, and E). In turn, these devices examine the address in the protocol data unit. If it is destined for the device, it copies and passes this packet to an upper layer protocol. If not, it simply ignores the packet.

Figure 1.4 is a more detailed view of the user access line. You can see that the user access line connects to the packet exchange (DCE) and the user device (DTE) through modems, service data units, multiplexers, etc. In turn, these devices are connected to the user device and the packet exchange through interchange circuits.

It is quite important to note once again that the DCE in Figure 1.3 can contain a fairly complex piece of equipment, such as a packet switch. From the context of the V Series Recommendations, a DCE is a modem, multiplexer, or a data service unit.

The Importance of Standards

Because the V Series Recommendations are used in many products and standards, it seems appropriate to include a brief discussion on why standards are so important to the data communications user. (If you are conversant with this subject, you can skip to the next section.)

Computer communication systems are used to support the transfer of data between two *end users*. An end user could be a person or even a computer

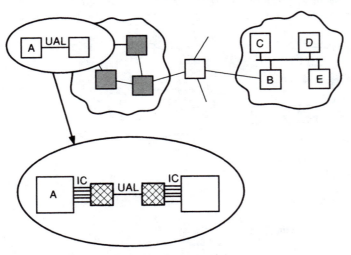

▨ = Modems, data service units, multiplexers, etc.

IC = Interchange circuits

Figure 1.4 User access line.

or terminal. Because the transfer is data (instead of voice or other images), the process is called *data communications* (some organizations, such as the ITU-T, prefer the term to be singular, as in *data communication*).

For this seemingly simple process to occur, the computers and their communications facilities must perform many actions. For example, the two machines must accept each other's communications characteristics (i.e., on a typical telephone line, the dial-up procedures that each machine uses to connect to the network). The machines must use the same signalling techniques, such as analog or digital signals. If the data are important, some means must be provided to assure that the data transfer through the network is successful and that both machines know that all data has been transferred without any problems.

These functions are not trivial, and because of their limited "intelligence," computers do not perform these tasks easily. Yet computers usually communicate without ambiguity if they are instructed (programmed) correctly, the communications signals between them are not distorted, and the computers understand the intent and meaning of each other's symbols.

This last requirement presents a formidable challenge because it implies a high degree of understanding and cooperation between the machines. In effect, it implies that the computers communicate with a common set of symbols and, equally important, have an unambiguous interpretation of these symbols. It implies the use of *standards*.

Standards not only ease the task of interfacing different computers (as in our example), but they also give the user more flexibility in equipment and software selection. In addition, the acceptance and use of a standard often leads to lower costs to consumers because a widely used standard can be mass-produced and perhaps implemented in very large-scale integrated (VLSI) chips.

In this regard, the V Series Recommendations have been instrumental in providing an effective platform for vendors to develop many of their data communications standards. Indeed, it is difficult to imagine how our modern world could function efficiently without the use of telecommunications standards.

The V Series and the OSI Model

The ITU-T V Series Recommendations have played a major role in fostering common data communications standards among different vendors and manufacturers. The recommendations have been accepted in Europe for many years. In the last few years, they have achieved worldwide use.

Many of the V Series Recommendations were published before the advent of the Open Systems Interconnection (OSI) Model. Nonetheless, the recent series have been written to conform to the OSI architecture, and some of the older series fit rather easily into the OSI structure. Because the

OSI is the keystone for some of the V Series Recommendations, this section provides a brief review of the OSI layers.

A review of the OSI layers

The seven OSI layers are depicted in Figure 1.5 and are summarized in this section. I also use this section to introduce some important attributes of the layers and the relationships of the layers to each other. Be aware that, from the context of the OSI Model, the vast majority of the V Series Recommendations operate at the physical level. However, as you shall see, some of the newer recommendations are intended for more sophisticated functions (such as error detection and correction and interworking with ISDN) and also provide functions at the data link level.

The lowest layer in the model is the *physical* layer. The functions within this layer are responsible for activating, maintaining, and deactivating a physical circuit between a DTE and a modem, multiplexer, or some other similar device. The layer also identifies the bits (as 0s or 1s).

Because this layer is concerned with the nature of the signals, it must be able to distinguish between different levels of voltages and the direction and intensities of currents. It is responsible for sending and receiving electromagnetic signals. It also is responsible for creating and interpreting the optical signals in optical fibers and for defining the cabling and wiring between machines (if any exist). In addition, it contains the specifications for the physical connectors used to attach cables and wires to the computers (which some people simply call *pins*).

The *data link* layer is responsible for the transfer of data over the communications channel (the channel also is called a link, a line, or a trunk). It provides for the synchronization of data to identify the bits coming from the physical layer. It also delineates each bit within each transmission. It ensures that the data arrive safely at the receiving computer or terminal. It provides for flow control to ensure that the computer does not become overburdened with too much data at any one time. One of its most impor-

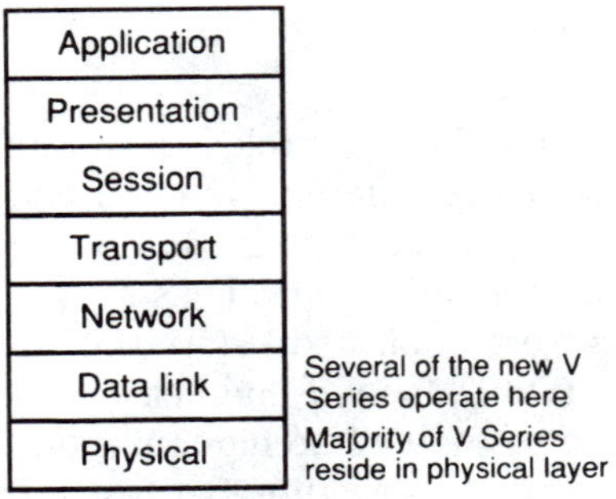

Application
Presentation
Session
Transport
Network
Data link
Physical

Several of the new V Series operate here

Majority of V Series reside in physical layer

Figure 1.5 The OSI Model and the placement of the V Series.

tant functions is to provide for the detection of transmission errors and to provide mechanisms to recover from problems that occurred on the link such as lost, duplicated, or damaged data. Several of the more recent V Series Recommendations operate at the data link layer.

The *network* layer specifies the interface of the user DTE *into* a network, as well as the interface of the DTEs with each other *through* a network. The network layer is the layer that is responsible for routing. In a packet switch network, this routing function is called *packet switching*, and in a circuit switch network, it is called *circuit switching*. The V Series Recommendations do not concern themselves with this layer or with any of the other upper layers. However, for the sake of continuity, I will briefly describe each of the remaining layers.

The *transport* layer provides the interface between the data communications network and the upper three layers. It is the layer that gives the user options in obtaining certain levels of quality (and cost) from the network itself (i.e., the network layer). It is designed to keep the user isolated from some of the physical and functional aspects of the network. It also provides for end-to-end integrity of the transfer of user data.

The *session* layer serves as the user interface into the transport service layer. The layer provides for an organized means to exchange data between end-user applications. The users can select the type of synchronization and control needed from this layer.

The *presentation* layer is used to ensure that user applications can communicate with each other, even though they might use different representations for their protocol data units (packets or messages). The layer is concerned with the preservation of the syntax of the data. For example, it can accept various data types (character, Boolean, integer) from the application layer and negotiate an acceptable syntax representation with another peer presentation layer, perhaps located in another computer.

The *application* layer is concerned with the support of an end-user application process. This layer contains service elements to support application processes such as job management, file transfers, electronic mail, and financial data exchanges.

Where the V Series reside in the OSI Model

Figure 1.5 shows the location of the V Series Recommendations in the OSI Model. All operate at the physical and data link layers. This approach keeps the V Series operations isolated from and transparent to the upper layers. For example, the application layer software has no knowledge of the activities of the V Series operations. Therefore, a change to the V Series layers does not affect the application layer software, nor does a change to the application layer software affect these lower layers.

Other aspects of OSI that pertain to the V Series

Other aspects of OSI that pertain to the V Series include horizontal and vertical communications and communications between layers.

Horizontal and vertical communications. OSI protocols allow interaction between functionally paired layers in different locations without affecting other layers. This concept aids in distributing the functions to the layers. In the majority of layered protocols, the data unit (such as a data link layer frame that is passed from one layer to another) usually is not altered, although the data unit contents can be examined and used to append additional data (trailers or headers) to the existing unit.

The relationships of the layers are shown in Figure 1.6. Each layer contains entities that exchange data with and provide functions to (horizontal communications) peer entities at other computers. For example, layer N in machine A communicates logically with layer N in machine B, and the $N+1$ layers in the two machines follow the same procedure. Entities in adjacent layers in the same computer interact through the common upper and lower boundaries (vertical communications) by passing parameters to define the interactions.

Communications between layers. Figure 1.7 depicts a layer providing a service or a set of services to users A and B. The users communicate with the service provider through an address or identifier that commonly is known as

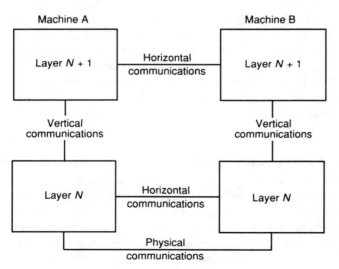

Figure 1.6 Horizontal and vertical communications.

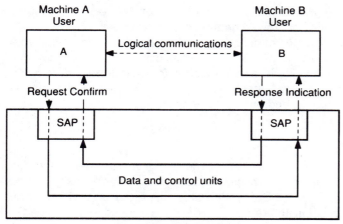

Figure 1.7 OSI primitives and service provisions.

the *service access point* (SAP). Through the use of four types of transactions, which are called *primitives* (request, indication, response, and confirm), the service provider coordinates and manages the communications process between the users; some sessions do not require all of the primitives:

- *Request.* Primitive by service user to invoke a function

- *Indication.* Primitive by service provider to invoke a function or indicate a function has been invoked at an SAP

- *Response.* Primitive by service user to complete a function that previously was invoked by an Indication at that SAP

- *Confirm.* Primitive by service provider to complete a function that previously was invoked by a Request at that SAP

ITU-T service definitions and protocol specifications

The ITU-T uses two OSI-based terms in describing several aspects of the V Series Recommendations:

- *Service definitions.* Define the services between the layers or entities within a layer, typically with the use of primitives

- *Protocol specifications.* Actions taken within the layer and between the peer layers as a result of the service definitions

The previous section explains the concepts of vertical and horizontal communications between layers. A convenient way to think of service definitions is that they provide vertical communication services. The protocol specifications provide horizontal communication services.

Many of the V Series Recommendations were published and installed in vendor products before the OSI Model was published. Therefore, not all of the V Series use the OSI Model as a foundation for their design. More recent Recommendations, such as V.42 bis, make use of the Model and define the use of service definitions and protocol specifications.

Summary

The V Series Recommendations are some of the most widely used data communications standards in the world today. The newer releases of these standards are based on the OSI Model. This chapter also provided a brief summary of the relationship of the V Series Recommendations to the OSI standards. The chapter emphasized that the use of standards is a preferable approach to that of allowing each manufacturer to develop a unique system.

With this information in mind, I now will examine the structure and organization of the V Series Recommendations. The reader should be aware of the subject matter in this chapter when reading the following chapters. On several occasions, I will refer to some ideas and terms that are introduced in this chapter.

Appendix 1A: The ITU-T

The ITU-T is a member of the International Telecommunications Union (ITU), a treaty organization that was formed in 1865. The ITU now is a specialized body within the United Nations. The former CCITT was formalized as part of ITU in 1956. ITU-T sponsors a number of recommendations that primarily deal with data communications networks, telephone switching standards, digital systems, and terminals. Each member country casts a vote on the ITU-T issues.

The ITU-T's Recommendations (also informally known as standards) are very widely used. Until recently, its specifications were republished every four years in a series of books that take up considerable space on a book shelf. The four-year period books are identified by the color of their covers. The colors are used in the following order: red, blue, white, green, orange, and yellow. The 1960 books were red; 1964, blue; 1968, white; 1972, green; 1976, orange; 1980, yellow; and, in 1984, once again red. The 1988 Blue Books consume about three feet of shelf space.

Changes made at Melbourne

The Melbourne meeting (in which the former CCITT approved the Blue Books) was a watershed for the ITU-T in that it realized that the increasing number of ITU-T Standards dictated a change to its organizational structure. First, the Study Groups were restructured.

Perhaps the most significant change that occurred at Melbourne was the administration of the ITU-T Recommendations. The ITU-T Recommendations now are published on a more frequent basis. The conventional four-year cycle has been eliminated, and the standards are published once a 70% or more approval vote is reached by the members. If the proposal is not accepted, it will be sent back to the Study Group.

This approach eliminated the 16,000-page publish-all-at-once mess that has resulted from the previous approach. It meant the end of the strict delineation of Yellow, Red, Blue Books, etc.

Since the Melbourne meeting, the ITU has reorganized and renamed the CCITT as the International Telecommunication Union-Telecommunication Standardization Sector (ITU-T). In effect, on February 28, 1993, the name "CCITT" no longer existed. The new organization is reflected in Figure 1A.1. The Study Groups (SGs) also were realigned, and their revised names and responsibilities are summarized in Table 1A.1. The members of the SGs are sent by their respective national telecommunications administrations to

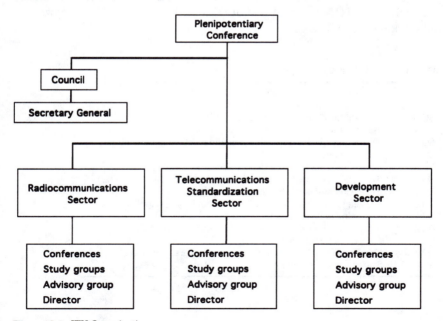

Figure 1A.1 ITU Organization.

participate in the SG activities. In 1993, 181 administrations were represented. However, other participation is allowed:

- Recognized operating agencies (ROAs), such as AT&T, MCI, and Sprint
- User organizations from industry, such as the airline industry
- Regional standards bodies, such as ETSI in Europe
- Manufacturers of telecommunications gear
- Scientific and industrial organizations (SIOs)

As stated earlier, the formal documents produced by the ITU-T are called *Recommendations*, although people are increasingly using the term *standards*, because many of them do become standards in many countries.

The Study Groups can be very large, sometimes involving hundreds of participants; therefore, SGs are divided into Working Parties, which can be divided further into Expert Groups, and Ad Hoc groups. The idea is for these bodies to "do the work" and submit (as much as possible) well-thought-out solutions and documents to the SGs. Each SG has a number of Raporteurs who are responsible for coordinating activity, and keeping the process going forward.

Table 1A.2 lists the ITU-T Series and a general description of their scope. You might find ITU-T Series A useful. It describes the ITU-T's organization and working procedures.

TABLE 1A.1 ITU-T Study Groups

SG 1	Service definition, operations, and quality of service
SG 2	Operation of network
SG 3	General tariff principles, includiing accounting
SG 4	Maintenance of all aspect of the network
SG5	Protection against environmental and electromagnetic effects
SG 6	Outside plant
SG 7	Data communications networks and OSI
SG 8	Terminal equipment for telematic services (fax, teletext, etc.)
SG 9	Television and sound transmissions
SG 10	Languages and methods for telecommunications applications
SG 11	Switching and signalling
SG 12	Transmission performance of networks and terminals
SG 13	General network aspects
SG 14	Transmission of data over telephone networks (modems)
SG 15	Systems and equipment

TABLE 1A.2 The ITU-T Recommendations

Series number	Series scope
A	ITU-T organization
B	Means of expression
C	General telecommunications statistics
D	General tariff principles
E	Telephone network and ISDN: operation, numbering, routing, and mobile service
F	Operations and quality of service (telegraph, telex, mobile, satellite, etc.)
G	Transmission systems and media
H	Transmission of nontelephone signals
I	ISDN
J	Transmission of sound program and television signals
K	Protection against interference
L	Construction, installation, and protection of cables and other elements of outside plants
M	Maintenance (international transmission systems, telephone circuits, telegraphy, facsimile, and leased circuits)
N	Maintenance of international sound program and television circuits
O	Specifications for measuring equipment
P	Telephone transmission quality
Q	Telephone switching and signalling
R	Telegraphy
S	Telegraph services terminal equipment
T	Terminal equipment and protocols for telematic services
U	Telegraph switching
V	Data communication over telephone networks
X	Public data networks
Z	Programming languages

2

Principal Operations
of the V Series

This chapter describes the principal operations of the V Series. These operations pertain primarily to the operations of the physical layer of the Open System Interconnection (OSI) Model. To briefly iterate our discussions in the previous chapter, physical layer protocols (or physical layer interfaces) are so named because the physical layer operations are concerned with the nature of the signals such as the different levels of voltages, the direction and intensities of currents, and the shaping of the physical signal (such as digital or analog signals). This layer also is responsible for defining the physical cabling and wiring between machines (if any exist).

Many of the V Series Recommendations describe the following physical layer operations:

- Description of the procedures for data transfer across the interface between the data terminal equipment (DTE) and the data circuit-terminating equipment (DCE) (in ITU-T terms, the *interchange circuits*)

- Provision for *control* of interchange circuits between the devices to govern how the *data* interchange circuits operate

- Description of clocking signals on specific interchange circuits to synchronize data flow and regulate the bit transfer rate between the DTE and the DCE

- Description of signals to synchronize data flow and regulate the bit transfer rate between the two DCEs

- Description of electrical ground

- Description of the mechanical connectors (such as pins, sockets, and plugs)

Most physical layer protocols describe four attributes of the interface: electrical, functional, mechanical, and procedural. The electrical attributes describe the voltage (or current) levels, the timing of the electrical signals, and all of the other electrical characteristics (capacitance, signal rise time, etc.). The functional attributes describe the functions to be performed by the interchange circuits at the physical interface. Many physical layer protocols classify these functions as control, timing, data, and ground. The mechanical attributes describe the dimensions of the connectors and the number of wires on the interface. Usually the data, signalling, and control wires are enclosed in one cover. The procedural attributes describe what the connectors must do and the sequence of events required to effect the data transfer across the interface.

Examples of Physical Layer Devices

The V Series Recommendations describe the operations of physical layer devices, such as the data circuit-terminating equipment (DCE) that was introduced in chapter 1. The DCE typically is a modem, multiplexer, or a digital service unit (also called a data service unit, a channel bank, or a channel service unit).

A modem is responsible for providing the translation and interface between the digital and analog worlds. The term *modem* is derived from the process of a local modem accepting digital bits from the local DTE and changing them into a form suitable for analog transmission (*mo*dulation) and from receiving the signal at the remote modem and transforming it back to its original digital representation (*dem*odulation) for the remote DTE.

Modems are designed around the use of a *carrier frequency*. This signal has the digital data stream superimposed on it at the transmitting end of the circuit. The receiver receives the carrier and demodulates the signal to derive the bits in the data stream. (The carrier frequency has the characteristics of the analog wave, which will be discussed later in this chapter.)

A short-haul or limited-distance modem (LDM) is used for transmissions of a few feet to approximately 20 miles. The distance is highly variable and depends on the operating speed, type of transmission path, and configuration of the telephone company facilities. Typical LDMs use pairs of wires or coaxial cables for the transmission media. Some can operate at relatively high data rates (19.2 kbit/s to 1 Mbit/s).

Line drivers are direct current (dc) machines and often are used in place of modems. Operating distances range from a few feet to several miles. The line driver usually is installed between two DTEs to replace the modems (on short distances). These machines are attractive because of their low price.

A null modem actually is an EIA-232 (which will be discussed later in this book) cable interface "pinned" to allow a direct connection between two devices when a modem is not required. Null modems provide no timing signals. Consequently, they are used with asynchronous devices that derive their timing from start/stop bits.

A modem eliminator is used in situations where two synchronous devices (such as terminals, printers, plotters) are in close proximity. The modem eliminator provides the interface and clocking for the devices.

Some people do not distinguish between a null modem and a modem eliminator. I use the term *null modem* to describe an interface for asynchronous devices and the term *modem eliminator* as an interface for synchronous devices. The exact use of the terms is somewhat irrelevant as long as the conversing parties use the same term to describe the same device.

Acoustic couplers are yet another alternative for physical layer signalling. These machines acoustically connect (with audible frequencies) the terminal to the analog facilities. The telephone handset is placed into the coupler's transmit/receive device for connection. Portable terminals might contain acoustic couplers. The technique is very simple and effective for low data rates. These couplers operated up to 1200 bit/s in the early 1990s, but products now are available for data rates of 2400 bit/s (typically, using the V.22 bis modem signaling conventions).

V Series Communications Media

The path for the transfer of data between DTEs and DCEs can take several different forms. This section describes the methods that are included in the ITU-T V Series Recommendations as well as their transmission characteristics.

Wire pairs

The early telephone communications systems used *open* wire pairs. They consist of uninsulated pairs of wires strung on poles about 125 feet apart. The air space between the wires provides insulation (isolation) from each other. Open wire systems have largely disappeared. They proved inadequate in large urban areas and were unattractive as well as cumbersome to install and maintain.

The early installations of wire pairs provided fairly good signal quality, but the two wires experienced mutual interference (crosstalk). Later, the pairs were twisted around each other to compensate for (and cancel) the effects of pair interference. These circuits are pervasive today and are called *twisted pair* cable (see Figure 2.1). They can be used to support both analog and digital signals and are assumed to be the media in many of the ITU-T V Series Recommendations.

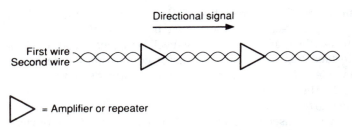

 = Amplifier or repeater

Figure 2.1 Twisted-pair cable.

Balanced and unbalanced twisted pairs

A two-wire system often insulates the conductors from each other in a cable. In addition to this insulation, the system has an outer insulation cable. At one time, the outer insulation was made of compositions such as lead. Modern insulations use various kinds of plastic or similar substances.

The twisting of each pair (in a multipair cable) is staggered. Radiated energy from the current flowing in one wire of the pair is largely canceled by the radiated energy of the current flowing back in the return wire of the same pair. This approach greatly reduces the effect of crosstalking signals between the wires. Moreover, each pair in the cable is less susceptible to external noise; the pair cancels out much of the noise because noise is coupled almost equally in each wire of the pair. This technique often is called *telephone twisted pair* (TTP).

In a balanced line, both wires carry current; the current in one wire is 180° out of phase with the current in the other wire. Both wires are above ground potential. In contrast, an unbalanced line carries the current on one wire, and the other wire is at ground potential.

Grounding is defined as a conducting circuit through which equipment is connected to the earth or to a conducting body that is at earth potential. *Potential* is a term to describe the differences of charge between two points on a circuit. Anything that is at ground potential has the same electrical charge as the earth. Ground potential describes anything that has the same charge (zero) as the earth ground.

Shielded and unshielded pairs

The bulk of TTP cable is unshielded copper wire. Unshielded wire has no metallic cover around it. These wires are inexpensive and offer reasonable performance over a long distance. However, they are quite susceptible to crosstalk and interference with external power sources.

Shielded twisted pairs (also called data-grade media) improve the resistance to crosstalk and external noise. The shield surrounds the wire with a metallic sheathing or braid. Several analyses reveal that a shielded pair system reduces crosstalk and improves noise resistance by a factor of 1000 or more. Attenuation of TTP is about 2.3 times more severe than shielded pairs.

Line Characteristics

The transmission line provides the media for the data exchange between stations. This section discusses the following characteristics of lines:

- Point-to-point and multipoint configurations
- Simplex, half-duplex, and duplex arrangements
- Switched and leased lines

Point-to-point and multipoint configurations

A point-to-point line connects two stations (see Figure 2.2a); a multipoint line has more than two stations attached (see Figure 2.2b). The selection of one of these configurations is dependent on several factors. First, a point-to-point arrangement might be the only viable choice if a prolonged, dedicated user-to-user session is necessary. Second, the traffic volume between two users might preclude the sharing of the line with other stations. Some high-speed computer-to-computer sessions require a point-to-point arrangement. Third, two users might be the maximum number involved in a communications session. A multidrop arrangement commonly is used in situations where low-speed terminals communicate with each other or a computer. The line is shared by the stations, thereby providing for its more efficient use.

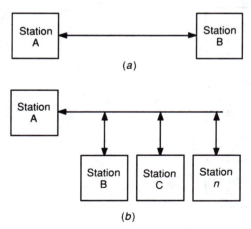

Figure 2.2 (a) Point-to-point; (b) multipoint configurations.

Multidrop lines require the use of more elaborate controls than do point-to-point lines. The stations on the multidrop path must be supervised to provide for the allocation and sharing of the line. Sessions must be interleaved, and priorities must be established for the more important sessions.

Simplex, half-duplex, and duplex arrangements

The terms *simplex*, *half-duplex*, and *duplex* often are subject to more than one interpretation. They should refer to the manner in which traffic flows across the line. Figure 2.3 illustrates the configurations for these arrangements.

A simplex transmission provides for the movement of traffic across the path in one direction only. The sender cannot receive, and the receiver cannot send. Commercial radio and television broadcasts are examples of simplex transmissions. The scheme is used in numerous applications but is not included in the V Series Recommendations.

Half-duplex transmission provides for movement of data across the line in both directions but in only one direction at a time. Human-operated keyboard terminals and workstations commonly use this approach. Typically, a message sent to the terminal requires the operator to read the message, enter an appropriate response, and send a reply. The terminal and the other station take turns using the line; the sending station waits for the response before sending another message. Half-duplex also is called *two-way alternate transmission* (TWA).

Duplex transmission (also called *full-duplex*) provides for simultaneous, two-way transmission between the stations. Multipoint lines frequently use this approach. For example, station A sends traffic to a central computer at

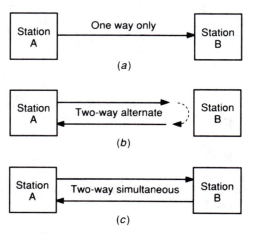

Figure 2.3 Data flow on the circuit. (*a*) Simplex; (*b*) half-duplex; (*c*) duplex.

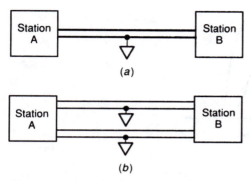

Figure 2.4 (*a*) Two-wire; (*b*) four-wire circuits.

the same time that the computer sends traffic to station B. Duplex transmission permits the interleaving of sessions and user data flow among several or many stations. Duplex also is called *two-way simultaneous transmission* (TWS).

Two-wire and four-wire arrangements

The telephone-type physical lines or paths are described as two- or four-wire circuits. Figure 2.4*a* depicts a two-wire configuration. This arrangement provides for two conductors, but only one is used for data exchange. The second conductor is a return channel or common ground to complete the electrical circuit. A four-wire circuit is shown in Figure 2.4*b*. In this case, four conductors are used to provide two data transmission paths and two return channels.

Split stream channels

A number of the V Series low- to medium-speed DCEs achieve full-duplex transmission on a two-wire circuit by dividing the bandwidth on the transmission wire (see Figure 2.5). This operation is called *split channel modulation* or *reverse channel modulation*. The technique uses two carriers, operating at different frequencies. These characters then are modulated by the data. Typically, the carrier is modulated up or down by approximately 100 Hz to represent 1s and 0s. As you will see, the V Series modems vary the carrier frequency by the amount of frequency shift.

The common approach for these machines to communicate is to designate one of them as the originate modem and the other one as the answer modem. These terms stem from the older Bell 103 modems that use the originate mode for a call that is originated by a device and the answer mode for a call that is answered by another device. The new V Series modems no

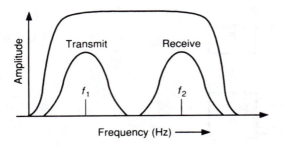

f = carrier frequency

Figure 2.5 Split stream.

longer use split stream channels but rely on a technique called *echo can-cellation*, which is described next.

Echo cancellation

Prior to the publication of the 1984 Red Book from the ITU-T, the V Series modems used frequency division modulation (FDM) to achieve full-duplex operations. Two channels were used to send and receive data simultane-ously. All the newer V Series modems now use echo cancellation for full du-plex operations.

Echo cancellation is a technique for handling echoes on the communica-tions path (the reception at the transmitter of the transmitted signals). It now also is used for obtaining full-duplex operations on a two-wire circuit. In conventional echo cancelers, a canceler is placed near the origin of the echo. A reference signal (Y) is sent to the canceler. In effect, this is a replica of the transmitted signal. The replica is inverted and stored as R_1. The echo (R) is returned along with the near-end talker signal. The canceler then uses the R_1 signal to subtract from the signal. In effect, the (R_1) is an in-verted replica of the transmitted voice pattern. By superimposing it onto the returning signal, the echo is canceled. The same concept can provide for full-duplex capabilities by the transmitter canceling its transmitted signal.

Switched and leased lines

The use of switched lines is well known to all of us because we use such fa-cilities to make telephone calls. The dial-up telephone uses the public tele-phone exchange. The switched line is a temporary connection between two sites for the duration of the call. A later call to the same site might use dif-ferent circuits and equipment in the telephone system. A leased line is a permanent connection between sites and does not require a user or com-puter to dial up the other user or computer.

Synchronizing Data Communications Components

The layers above the physical layer assume that the devices on the link already are "physically" connected and communicating with each other. This means that the physical layer must be operating properly for the data to be transmitted across the communications channel.

For machines to communicate, they first must notify each other that they are about to transmit data. Second, once they have begun the communications process, they must provide a method to keep both devices aware of the ongoing transmissions. To address the first point, consider that a transmitter, such as a terminal or a computer, must transmit its signal so that the receiving device, such as a modem or digital service unit, knows when to search for and recognize the data as they arrive. In essence, the receiver must know the exact time that each binary 1 and 0 is propagating across the wire (or some other type of media). This requirement means that a mutual time base or a "common clock" is needed between the receiving and transmitting devices.

The transmitting machine first forwards to the receiving machine an indication that it is sending data—something like a human saying "hello." If the transmitter sends the bits across the channel without prior notice, the receiver probably will not have sufficient time to adjust itself to the incoming bit stream. In such an event, the first few bits of the transmission would be lost, perhaps rendering the entire transmission useless. Moreover, the receiver might not be able to "train" itself onto the transmission if it does not detect the first part of the signal.

This process is part of a communications protocol and generally is referred to as *synchronization*. Connections of short distances between machines (e.g., between a computer and a modem) often use a *separate channel* to provide the synchronization. This line transmits a signal that is turned on and off or varied in accordance with preestablished conventions. As the clocking signal on this line changes, it notifies the receiving device that it is to examine the data line at a specific time. It also might adjust the receiver's sampling steps to enable the receiver to stay accurately aligned on each incoming data bit. Thus, clocking signals perform two valuable functions: they synchronize the receiver onto the transmission before the data actually arrive and keep the receiver synchronized with the incoming data bits.

The best approach for achieving synchronization is to use a code with frequent signal-level transitions on the channel. The transitions delineate the binary data cells (1s and 0s) at the receiver, and sampling logic continuously examines the state transmissions to detect the bits. Receiver sampling usually occurs at a rate that is higher than the data rate to define the bit cells more precisely.

Many of the V Series Recommendations specify the operations for a *scrambler*. It operates on the communications link between the modems (not on the interchange circuits between the DTE and the modem) and is used to guarantee that the incoming data stream will have sufficient transitions for the receiver to recover the signal accurately. The scrambler at the transmitting modem is used for randomizing the signal. The signal is translated back to its original form at the receiver by yet another scrambler.

Encoding and decoding

To gain a better appreciation of how machines communicate with each other, we need to go one step further and examine the actual signals on the interchange circuits. Figure 2.6 shows how signals are used to achieve synchronization. Note that the term *code* is used here in a different context than with the EBCDIC/ASCII/IA5 codes. Figure 2.6*a* shows the nonreturn-to-zero code (NRZ). The signal level remains stable throughout the bit cell. In this case, the level remains at a negative voltage for a 1 bit and goes to a zero voltage for a 0 bit. (Opposite voltages also are used in some devices.)

NRZ is a widely used data communications scheme because of its relative simplicity and low cost. The NRZ code also makes very efficient use of the channel because it represents a bit with each signal change (in some systems, no signal change actually can represent a bit). However, it does not have very good self-clocking capabilities because a long series of continuous 1s or 0s would not create a signal state transition on the channel. As a consequence, the receiver's clock could possibly "drift" from the incoming signal and sample the line at an incorrect time. The transmitter and the receiver actually might lose synchronization with each other.

The return-to-zero code (RZ) entails the changing of the signal at least once in every bit cell. This scheme is illustrated in Figure 2.6*b*. Because RZ codes provide a transition in every bit cell, they have very good synchronization characteristics. A primary disadvantage of RZ is the requirement for two signal transitions for each bit. Consequently, an RZ code requires at least twice the channel capacity (bandwidth) of a conventional NRZ code

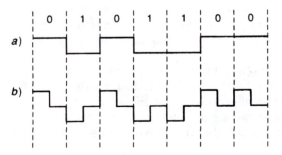

Figure 2.6 Digital signals on the interchange circuit. (*a*) NRZ; (*b*) RZ.

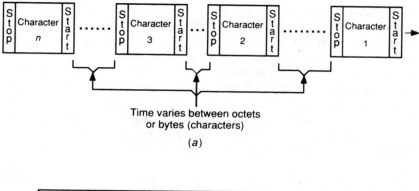

Time varies between octets
or bytes (characters)

(*a*)

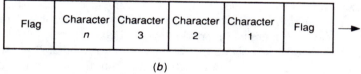

(*b*)

Figure 2.7   Transmission formats. (*a*) Asynchronous; (*b*) synchronous.

and is a more expensive approach. This type of code is used in some of the more sophisticated systems such as local area networks (LANs), lightwave technologies, and other high-speed channels.

Asynchronous and Synchronous Transmission

The vast majority of computers and terminals implement the NRZ code in their communications interfaces—for example, the interface between the DTE and the DCE (modem)—and the V Series Recommendations specify an NRZ signal on most of the DTE-DCE interfaces. (This Recommendation is described in chapter 4 along with V.28.) Consequently, clocking becomes a major consideration with these devices.

Two data formatting conventions are used to help achieve synchronization and are illustrated in Figure 2.7. The first approach is called *asynchronous formatting*. With this approach, each data byte or octet (character) has start and stop bits (i.e., synchronizing signals) placed around it. The purposes of these signals are to alert the receiver that data are arriving and to give the receiver sufficient time to perform certain timing functions before the next character arrives. The start and stop bits really are nothing more than unique and specific signals that are recognized by the receiving device.

Asynchronous transmission is widely used because the interfaces in the DTEs and DCEs are relatively inexpensive. For example, most personal computers use asynchronous interfaces. Because the synchronization occurs between the transmitting and receiving devices on a character-by-character basis, some allowance can be made for inaccuracies because the inaccuracy can be corrected with the next arriving character. In other words, a "looser" timing tolerance is allowed, which translates to lower component costs.

A more sophisticated process is *synchronous transmission*. It uses separate clocking channels or a self-clocking code. Synchronous formats eliminate the intermittent start/stop signals around each character and provide signals that precede and sometimes follow the user data stream. The preliminary signals usually are called synchronization (sync) bytes, flags, or preambles. Their principal function is to alert the receiver of incoming user data. This process is called *framing*.

A long synchronous data message without intermittent start/stop bits presents timing problems because the receiver might drift from the signal. The previous section explored two ways to deal with this problem: provide a separate clocking channel or provide a signal code that is self-clocking, such as the RZ code, or a signal that otherwise is scrambled. The latter approach allows the receiver to derive its timing from the line transitions by synchronizing a local sample clock onto the incoming signals.

Synchronization and Clocking in the V Series Recommendations

Synchronization and clocking are important functions of the physical layer. As discussed in the previous section, asynchronous and synchronous systems use different synchronization procedures. You learned that asynchronous transmission provides timing signals by the start/stop bits, and I also introduced the idea of clocking for synchronous systems. Synchronous transmission provides timing signals by:

- A separate clocking line
- Embedding the clocking signal in the data stream by randomizing the signal with a scrambler

Separate clocking line

A separate clocking line is a widely used technique in the V Series Recommendations for short-distance connections between the DTE and the DCE on the interchange circuits. In addition to the data line, another line transmits an associated timing signal, which is used to "clock" the data into the receiver. The clocking line notifies the receiver when a bit is arriving. However, the transmitting station does not always provide the clocking signal. Many configurations use the receiver (typically the modem) to provide the clocking signal to the transmitting station (the DTE). This signal from the receiver dictates the specific time for the transmitter to send data. This feature also is supported in the V Series Recommendations.

A separate clocking line is a common technique for synchronous inter-faces between terminals and/or computers and their associated DCEs, such as modems and multiplexers. However, a separate clocking channel is not practical under certain conditions:

- Longer distances make a separate wire prohibitively expensive.

- Longer distances also increase the probability that the clocking line will lose its synchronization with the data line, because each line has its own unique transmission characteristics.

- The telephone network does not provide clocking lines for a typical data subscriber because the lines are designed for voice.

Scrambler

The second approach is to embed the clocking signal in the data and have the receiver extract the clock from the received data stream. The receiver uses relatively simple circuitry for the clock extraction. To embed the clock-ing signal, the data bits are encoded at the transmitter to provide frequent transitions on the channel. As stated before, this operation is accomplished with the scrambler and is used by a number of the V Series modems.

In its simplest form, scrambling enables the modems to stay in sync by ensuring that the signal (combinations of 1s and 0s) changes frequently. Therefore, the newer V Series modems use a predefined algorithm to code and decode respectively the transmitted and received signal.

As an example of the scrambling and unscrambling process, V.32 bis uses a scrambler for each direction of transmission. A call mode generating poly-nomial (GPC) of $1 + x^{-18} + x^{-23}$ and an answer mode generating polynomial of $1 + x^{-5} + x^{-23}$ are used for each direction.

For the transmitting end, the user message is divided by the generating polynomial. The output of the scrambler is the coefficients of the quotients of this division, in descending order. For the receiving end, the user message is multiplied by the generating polynomial to recover the message sequence.

Analog and Digital Signals

The data communications systems that use the telephone line must cope with the fact that the majority of telephone local loops to homes and offices are designed and constructed around human speech. Those signals exhibit the analog waveform characteristics discussed in the next section. *Analog* means that a periodic signal gradually changes its amplitude levels or cur-rent strength. *Periodic* means that the wave variation from crest to crest is just like the other preceding waves (see Figure 2.8).

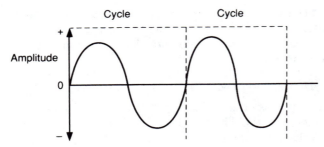

Figure 2.8 Analog signal.

However, computers, printers, terminals, and other data communications devices are digital devices; numbers and other symbols are represented by discrete signals. Typically, these symbols take the form of binary (1 and 0) values by abrupt changes in voltage levels or current flow (see Figure 2.9).

It now should be readily apparent why you must understand the nature and characteristics of the analog waveform. It is used to "carry" data across an analog communications link to and from computers and terminals.

The Waveform

When you speak into a telephone, the telephone handset transforms the physical speech waveform into an electrical waveform. Both waves have very similar characteristics. For example, the various heights of the sound wave are translated by the telephone into signals of continuously variable electrical voltages, or currents.

The voice waveform that is spoken into a telephone creates an electrical alternating current. The alternating voltage reverses its polarity, which produces current that reverses its direction.

Using a simple electrical generator as an example, the amount of voltage depends on the position of the loop conductors in the magnetic field. As the conductor rotates counterclockwise through the field, at ¼ turn, it produces the maximum positive voltage; at ½ turn, zero voltage; at ¾ turn, the maximum negative voltage; and at the completion of one turn, zero voltage.

The relationship of the generator rotation and the analog waveform is as follows (see also Figure 2.10):

A: 0° = 0 turn
B: 90° = ¼ turn
C: 180° = ½ turn
D: 270° = ¾ turn

The full revolution creates an electrical waveform of varying voltages and alternate directions in current flow (i.e., an alternating current). The wave-

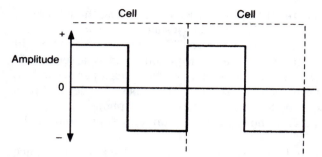

Figure 2.9 Digital signal.

length is shown as the variations between two successive points having the same voltage value and varying in the same direction of current flow.

Alternating current (ac) can be fed to a transformer that changes its voltage levels up or down to represent the waveform. This technique allows large amounts of power to be transmitted over high-capacity lines, yet transformed down for use in homes and offices. This characteristic is the reason that ac is used in many power systems and communications systems.

Direct current is the type of current created by batteries. It produces a steady voltage that produces a steady current in one direction only. Even though it cannot be stepped up or down with transformers, it is a voltage source for many communications systems.

Cycles, frequency, and period

Several terms now are introduced to compare the voice signal to its electrical counterpart and to the digital signal stored and used in a computer. The key terms are *cycle*, *frequency*, and *period*.

A: $0° = 0$ turn
B: $90° = 1/4$ turn
C: $180° = 1/2$ turn
D: $270° = 3/4$ turn

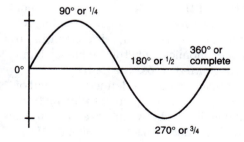

Figure 2.10 Phases of a signal.

The wavelength measures the cycle of the wave; that is, the interval of space or time in which the waveform reaches a successive point. The cycle describes a complete oscillation of the wave. The number of oscillations of the acoustical or electrical wave in a given period (usually a second) is called the frequency (f). Frequency is expressed in cycles per second or, more commonly, hertz. Frequency describes the number of cycles that pass a given point in one second (for example, an ear, a telephone mouthpiece, or a receiver in a computer). The signal travels one wavelength during the time of one cycle.

The time that is required for the signal to be transmitted over a distance of one wavelength is called the period (T). The period describes the duration of the cycle and is a function of the frequency:

$$T = 1/f$$

Also, frequency is the reciprocal of the period:

$$f = 1/T$$

Bandwidth

The analog voice signal is not made up of one unique frequency. Rather, the signal on a communications line consists of waveforms of many different frequencies. The particular mix of these frequencies is what determines the sound of a person's voice. Many phenomena manifest themselves as a combination of different frequencies. The colors in the rainbow, for instance, are combinations of many different lightwave frequencies; musical sounds consist of different acoustic frequencies that are interpreted as higher or lower pitch. These phenomena consist of a range or band of frequencies called the bandwidth, stated as:

$$BW = f_1 - f_2$$

where BW is the bandwidth, f_1 is the highest frequency, and f_2 is the lowest frequency.

As examples, a piano can produce a wide range of frequencies ranging from about 30 Hz (low notes) to over 4200 Hz (high notes). Its bandwidth is from 30 to 4200 Hz. The human ear can detect sounds over a range of frequencies from around 30 to 18,000 Hz, but the telephone system does not transmit this band of frequencies. The full range is not needed to interpret the voice signal at the receiver because most of the energy is concentrated between 300 and 3100 Hz. Because of economics, only the frequency band of approximately 200 to 3500 Hz is transmitted across the path.

The so-called voice-band (or voice-grade) channel is defined as a band of 4000 Hz. This means that the channel consists of frequencies ranging from 0 to 4000 Hz. The speech signal is band limited to between 200 and 3500 Hz. For purposes of convenience and brevity, industry uses the value 3 kHz as the bandwidth for a voice-band channel. The other frequencies on both sides of the speech signal allow for guard bands, which lessen interference between the channels that are placed on the same physical media such as a wire or cable.

With very few exceptions, the V Series Recommendations define data communications operations across a telephone or telephone-type voice-band channel.

Broadband and baseband signals

Signals usually are categorized as either broadband or baseband. A broadband signal is identified by the following characteristics:

- Uses analog wave forms
- Has a large bandwidth (typically in the megahertz to gigahertz range)
- Uses analog modulation
- Often uses frequency-division multiplexing for channel sharing

A baseband signal is identified by the following characteristics:

- Uses digital signals (voltage shifts)
- Has limited bandwidth
- Does not use modulation
- Can use time-division multiplexing for channel sharing

Many people use the term *baseband* to describe an unmodulated signal. A baseband signal can be used to modulate an analog carrier signal, but the carrier need not be a broadband carrier; it can be a voice-band carrier. Obviously, this signal is not considered a broadband signal.

Filters

It often is desirable to separate the frequencies that operate on a circuit. To emphasize the importance of frequency separation, consider these examples. The lower and higher frequencies on a voice-grade channel are separated and blocked for purposes of economics and performance. Radio signals also require channel separation, and many signals are separated from each other through guard bands to prevent signal interference.

The most common approach is the use of a filter to select or reject a frequency or a group of frequencies. For example, a high-pass filter could pass a 10-kHz input signal but block a 100-Hz signal. Conversely, the low-pass filter could perform the opposite service.

Generally, data communications systems make use of four kinds of filters:

- High-pass filter: Attenuates lower-frequency signals and passes higher-frequency signals
- Low-pass filter: Passes lower-frequency signals and attenuates higher-frequency signals
- Band-pass filter: Attenuates frequencies above and below a specific level and passes the signals within the band
- Band-stop filter: Passes signals above and below a specific band and attenuates the signals within the specified frequency band

The Decibel

The engineers who design and maintain communications systems must be concerned with the quality and strength of the signals in the system. Typically, a communications link consists of a number of different components such as amplifiers, communications lines, and switches. Each component will introduce signal loss or gain into the signal. These losses and gains are described as a ratio of power into and out of the component:

$$Power\ ratio = power\ out/power\ in$$

A tandem link (in which several components connect the two communicating devices) necessitates a calculation of loss or gain at each component and the multiplication of these ratios together. For example, assume five components introduce the following losses or gains:

$$1/2 \times 1/10 \times 1/50 \times 1/50 \times 10000/1 = 0.2$$

Thus, the signal is twice as strong at the end of the tandem link as it was at the beginning.

This multiplication process can be tedious, and the numbers can be very small or very large. Instead of multiplying the numbers together, the same result can be obtained by adding the logarithms of the numbers. Today, the standard practice is to use the logarithms of the ratios rather than the ratios themselves.

The term decibel (dB) is used in communications to express the ratio of two values. The values can represent power, voltage, current, or sound levels. It should be emphasized that the decibel is a ratio and not an absolute value, expresses a logarithmic relationship and not a linear one, and can be used to indicate either a gain or a loss. The logarithm is useful because a signal's strength falls off logarithmically as it passes through a cable.

A decibel is 10 times the logarithm (in base 10) of the ratio:

$$dB = 10\log_{10}P_1/P_2$$

where dB is the number of decibels, $\log_{10}$ is the logarithm to the base 10, P_1 is one value of the power, and P_2 is the comparison of value of the power.

Decibels often are used to measure the gain or loss of a communications signal. These measurements are quite valuable for testing the quality of lines and determining noise and signal losses, all of which must be known to design a communications system. It is a very useful unit because it can be added or subtracted as a signal is cascaded through a communications link. For example, if a line introduces 1 dB of loss in a span of 1 mile, a 3-mile length will produce a loss of 3 dB. If the line is connected to an amplifier with a gain of 10 dB, the total gain is 7 dB.

Suppose a communications line is tested at the sending and receiving ends. The P_1/P_2 ratio yields a reduction of the signal power from the sending to receiving end by a ratio of 200:1. The signal experiences a 23-dB loss ($23 = 10\log_{10}200$). The log calculations are readily available from tables published in math books.

The decibel often is used to describe the level of noise on the circuit by a signal-to-noise ratio. As Table 2.1 shows, 0 dB is equivalent to a 1:1 ratio of the signal to noise.

Because the decibel is a ratio and not an absolute unit, it is meaningless to use if you have no reference level by which to apply the ratio. For example, a 30-dB increase of 1 W of power is considerably different from a 30-dB increase in 1 mW of power. It is common practice to use a reference level of a watt or a milliwatt.

The decibel-watt (dBW) is employed for microwave systems. The measurement is made to a reference and expressed as:

$$\text{Power (dBW)} = 10\log_{10}\text{power (W)}/1\ \text{W}$$

**TABLE 2.1 Decibels
and Signal-to-Noise Ratios**

Decibels (dB)	Signal-to-noise ratios
0	1:1
+3	2:1
+6	4:1
+9	8:1
+10	10:1
+13	20:1
+16	40:1
+19	80:1
+20	100:1
+23	200:1
+26	400:1
+29	800:1
+30	1000:1
+33	2000:1
+36	4000:1
+39	8000:1
+40	10,000:1

The dBm is used as a relative power measurement in which the reference power is 1 mW (0.001 W):

$$dBm = 10\log_{10} P/0.001$$

where P is the signal power in mW.

This approach allows measurements to be taken in relation to a standard. A signal of a known power level is inserted at one end and measured at the other. A 0-dBm reading means 1 mW.

Echoes

Almost everyone who has used a telephone has experienced the problem of echoes during a conversation. Echoes are caused by the changes in impedances in communications circuits. (*Impedance* is the combined effect of inductance, resistance, and capacitance of a signal at a particular frequency.) For example, connecting two wires of different gauges could create an impedance mismatch. Echoes also are caused by circuit junctions that erroneously allow a portion of the signal to find its way into the return side of a four-wire circuit.

An echo often is not noticed. The feedback on a short-distance circuit happens so quickly that it is not perceptible. Generally, an echo with a delay of greater than 45 ms (0.045 s) presents problems. For this reason, long-distance lines and satellite links employ techniques to reduce the strength of the echo or eliminate it completely. The V Series Recommendations contain a number of measures to deal with echo.

Echo suppressors are used by telephone companies to filter out the unwanted signals that are echoed back to a transmitter. A suppressor can be placed on the return circuit to suppress the echoes. The suppressor determines which signal is on the line from a talker and inserts a very high loss (35 dB or more) in the opposite direction of the speaker.

Echo suppressors cannot be used for data transmission over a voice line. The speech detector is designed to detect speech signals. Moreover, the delay in reversing the activation of the suppressors at each end often causes the clipping of the first part of a signal. It usually does not present serious problems in a voice transmission but can cause distortions in a data transmission.

To use the V Series modems, the echo suppressors must be disabled for data transmissions. This is accomplished by the transmission of a 2100-Hz signal for approximately 400 ms (a 2225-Hz signal is used with the older Bell 103 standard).

Echo cancellation is another technique for handling echoes and has replaced many of the echo suppressors. The V Series specify echo cancellation as well for a number of full-duplex DCEs.

An echo canceler operates in the following manner. A canceler is placed near the origin of the echo and a reference signal Y is sent to the canceler.

In effect, this is a replica of the transmitted signal. The replica is inverted and stored as R_1. The echo R is returned along with the near-end signal. The canceler then uses the R_1 signal to subtract from the signal. In effect, each modem learns about its signals and the characteristics of the reflected signals. It then subtracts its echoes from a received signal.

The Interface at the DTE

Most communications interfaces use an input-output (I/O) interface chip called the universal asynchronous receiver/transmitter (UART) or the universal synchronous/asynchronous receiver/transmitter (USART). This large-scale integrated (LSI) device performs the following functions:

- Accepts parallel data from the DTE bus and converts them to serial data for the communications link

- Accepts serial data from the link and converts them to a parallel form for the DTE bus

- Performs some limited problem detection (parity, no stop bit, character overrun)

- Provides for transmitter and receiver clocks

- Detects a byte or block of data from the link or from the DTE

- Selects a number of stop bits for an asynchronous system

- Selects a number of bits per character

- Selects odd or even parity

The UART usually is packaged in a 40-pin dual inline package, or DIP (see Table 2.2), and the pins are inserted into a board that becomes part of the DTE interface logic. The device is used by both the DTE and DCE to control the communications process. The pins are set to a high or low state to provide the functional signals described in Table 2.2.

A simple USART interface is shown in Figure 2.11. I will use this figure to examine how data are transferred through the interface. Data are transferred across the physical layer connector (the input interchange circuit) from the modem and communications link to an 8-bit UART or USART input buffer. When the buffer is filled, the data are moved across a parallel bus to the DTE memory for further processing.

On the transmit side, data are transferred after the DTE has filled the output buffer and sent a control signal to instruct the chip to send the data onto the link (the output interchange circuit).

The address selection logic allows the host to select the individual port and the registers at the port for either reading or writing. The interrupt control logic allows the host to service an interrupt request by the USART or UART chip.

TABLE 2.2 UART Pin Functions

Pin number	Name	Mnemonic	Functions
1	—	—	Power
2	—	—	Power
3	Ground	GND	Electrical ground
4	Received Data Enable	RDE	Gates the received data onto pins 5 through 12
5–12	Data Output	RD	Data out
13	Parity Error	PE	Parity error indicator
14	Framing Error	FE	No valid stop bit
15	Overrun	OR	Data overrun in the buffer
16	Status Word Enable	SWE	Places PE, OR, TBMT, FE, and DA on output lines
17	Receiver Clock	RCP	Input line for an external clock
18	Reset Data Available	RDA	Resets DA
19	Received Data Available	DA	Data character has been received
20	Serial Input	SI	Incoming data
21	External Reset	XR	Resets UART chip
22	Transmit Buffer Empty	TBMT	Transmit register can be loaded with character
23	Data Strobe	DS	Loads data into transmit register
24	End of character	EOC	Goes low when last bit of character is transmitted
25	Serial Output	SO	Outgoing data
26–33	Data Input	DB	Data in
34	Control Strobe	CS	Places certain information into holding register
35	No Parity	NP	No parity checks
36	Two Stop Bits	SB	Indicates number of stop bits
37, 38	Number of Bits	NB	Number of bits in character
39	Parity Select	POE	Indicates type of parity
40	Transmitter clock	TCP	Input line for external clock

The registers are set and/or examined by the host I/O program to manage the communications process. Each bit in the registers can be individually programmed to configure the communications process to suit the user's need.

Modems and Modulation

Digital signals are used in computer systems to represent binary numbers. Yet analog signals are found in most voice-oriented telephone systems, such as those systems used by the V Series Recommendations. It is common practice today to convert the digital signals to analog signals for transmission across the communications channel, and the V Series modems (which are described in chapters 4 and 5) are designed to perform this operation.

Three basic methods of digital-to-analog modulation exist. Some of the higher-speed modems use more than one of the methods. Each method impresses the data on a carrier signal, which is altered to carry the properties of the digital data stream. The three methods are amplitude modulation (AM), frequency modulation (FM), and phase modulation (PM). See Figure 2.12.

Modulation rate (baud) and bit rate (bit/s)

The carrier signal typically is changed as often as possible because each change represents binary data and because frequent changes increase the bit transmission rate. The change rate is called the *signalling rate, modulation rate,* or *baud.*

The terms *baud* and *bit/s* often are used incorrectly to convey the same meaning. In this section, the two terms will be defined. It will be evident that they are indeed different, except when applied to low-speed devices. (*Bit rate* is an informal and widely used term to describe the transfer rate of a system in bits per second.)

Multilevel modulation

Baud and bit rate are equal only if 1 bit is represented with each signal change (which also is called the *signalling interval*). Such is the case with the lower-speed V Series modems, typically in the range of 600 bit/s and less. Higher-speed modems use multilevel modulation in which more than 1 bit is represented with each baud. The baud is calculated as:

$$B = 1/T$$

where B is the baud and T is the length of signalling interval (in time).

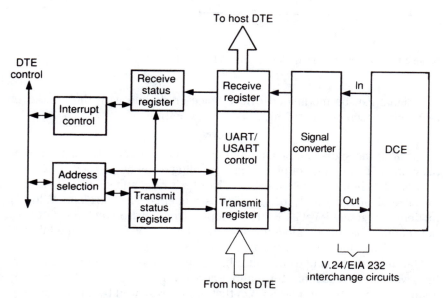

Figure 2.11 The DTE interface.

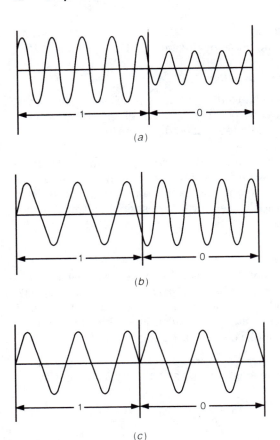

Figure 2.12 Modulation techniques. (*a*) AM; (*b*) FM; (*c*) PM.

A signal can be modulated over a voice-grade line with the multilevel technique based on the following general definition:

$$R = D/N$$

where R is the signalling or modulation rate in baud, D is the data rate in bit/s, and N is the number of bits per signalling element.

Table 2.3 shows the relationship of baud, bit rate, and signalling interval for several of the widely used V Series modems. The V.21 and V.26 bis modems are single-level DCEs; they employ a modulation technique of 1 bit per baud. The other modems are multilevel devices; the modulation schemes carry more than 1 bit per baud.

The multilevel modems permit a higher bit rate on the communications link, which achieves more throughput across a given bandwidth. However, a limit exists on the number of bits that are represented by each signalling interval because multilevel transmission is more expensive and more subject to error. The closer differences between the signal levels are more difficult to detect and more easily distorted. The higher-speed modems now encode 4 bits per baud.

If a limit exists on the number of bits per signalling interval, why not increase the modulation rate? For example, if the V.32 modulation rate of 2400 baud were increased to 4800 baud, the transmission rate could be increased to 19,200 bit/s. Such rates now are possible, but it is instructive to examine the column labeled signalling interval in Table 2.3. The higher-speed modem uses a signalling interval that is shorter than the lower-speed modems. For example, the V.21 interval is 0.003333 s and the V.29 interval is only 0.000416 s. The shorter signalling intervals are "less rugged"; they are more easily distorted on the communications link than the low-baud modems. Moreover, the higher signalling rates introduce other signals (more harmonics) and thus require more bandwidth.

For bandwidth-limited channels, multilevel transmission is achieved by applying the following formula:

$$R = \log_2 L \, 1/T$$

where R is the data rate in bit/s, L is the number of encoding levels (bits per baud), and T is the length of signalling interval.

Amplitude modulation

AM modems alter the carrier signal amplitude in accordance with the modulating digital bit stream (see Figure 2.12a). The frequency and phase of the carrier are held constant, and the amplitude is raised or lowered to represent a binary value of 0 or 1. In its simplest form, the carrier signal can be switched on or off to represent the binary state.

AM often is not used by itself because of transmission power problems and its sensitivity to distortion. However, it commonly is used with PM to yield a method superior to either FM or PM.

TABLE 2.3 Comparison of Bit, Baud, and Signalling Rates

CCIT-T specification	Modulation rate (baud)	Signalling interval (1/baud)	Bits per baud	Bit rate (bit/s)
V.21	300	0.003333	1	300
V.22 bis	600	0.00166	4	2400
V.26 bis	1200	0.000833	1	1200
V.26 ter	1200	0.000833	2	2400
V.27 ter	1600	0.000625	3	4800
V.29	2400	0.000416	3	7200
V.32	2400	0.000416	4	9600

The AM signal is represented as:

$$S(t) = A \cos(2\pi f_c t + \theta_c) \quad \text{binary 1}$$
$$S(t) = 0 \qquad\qquad\qquad\quad \text{binary 0}$$

where $S(t)$ is the value of carrier at time t, A is the maximum amplitude of carrier voltage, f_c is the carrier frequency, and θ_c is the carrier phase.

This approach also shows that a binary 0 is represented by no carrier, which is called *off/on keying* or *amplitude shift keying* (ASK).

The conventional AM data signals are detected at the receiver by envelope detection. The signal is rectified at the receiver and smoothed to obtain its envelope. This approach does not require the use of a reference carrier. Consequently, ASK is a relatively inexpensive process (the use of a carrier reference is a more complex and expensive undertaking). However, envelope detection does require both sidebands for accurate detection.

The off/on keying is a useful approach because it is simple. Nonetheless, ASK makes inefficient use of transmission power because the binary 0 signal is not the exact negative of the binary 1. The best use of power is achieved by having one signal exhibit the exact opposite polarity of the other signal.

The problem is overcome with phase reversal keying (PRK). The idea is to produce two identical signals with 180° phase reversal—hence the name PRK. It is obvious that PRK uses phase changes, but it still is considered a special form of AM.

PRK signals are detected at the receiver by the use of a coherent (or homodyne) carrier reference. A local receiver carrier is synchronized with the phase of the transmitter. With off/on AM systems, no information is contained in the phase. With PRK, the phase contains all of the data and synchronization information.

The use of AM for data communications has decreased because a multilevel scheme requires the use of several to many signal levels. As the number of signal levels increases, the distance between them decreases. The signal level's distance between the single and multiple level systems is quite close. AM transmitters often "saturate" these narrow distances. In some systems, they must be used at less than maximum power to diminish the saturation problem.

Thus, AM modems must be designed with sufficiently long signalling intervals (low baud) to keep the signal on the channel long enough to withstand noise and to be detected at the receiver and with sufficient distances between the AM levels to allow accurate detection and to diminish saturation.

Because of these and other problems, the vast majority of the V Series modems use other modulation techniques.

Frequency modulation

Figure 2.12b illustrates FM. This method changes the frequency of the carrier in accordance with the digital bit stream. The amplitude is held constant. In its simplest form, a binary 1 is represented by one frequency, and a binary 0 by another.

Several variations of FM modems are available. The most common FM modem is the frequency shift key (FSK) modem that uses four frequencies within the 3-kHz telephone line bandwidth. The FSK modem transmits 1070- and 1270-Hz signals to represent a binary 0 (space) and binary 1 (mark), respectively. It receives 2025- and 2225-Hz signals as a binary 0 (space) and binary 1 (mark). This approach allows full-duplex transmission over a two-wire voice-grade telephone line. Frequency shift keying is expressed as:

$$S(t) = A \cos(2\pi f_1 t + \theta_c) \quad \text{binary 1}$$
$$S(t) = A \cos(2\pi f_2 t + \theta_c) \quad \text{binary 0}$$

FSK is used for low-speed modems (up to 1200 bit/s) because it is relatively inexpensive and simple. Many personal computers use FSK for communications over the telephone network. FSK also is used for radio transmission in the high-frequency ranges (3 to 30 MHz), and some LANs employ FSK on broadband coaxial cables. However, its use on voice-grade lines is decreasing as ITU-T publishes more recommendations on PM, and more manufacturers implement DCEs with PM techniques.

Phase modulation

Previous discussions of the analog signal describe how a cycle is represented with phase markings to indicate the point to which the oscillating wave has advanced in its cycle. PM modems alter the phase of the signal to represent a 1 or 0 (see Figure 2.12c).

The PM method also is called *phase shift key* (PSK). A common approach to PSK is to compare the phase of the current signal state to the previous signal state, which is known as *differential PSK* (DPSK). This technique uses bandwidth more efficiently than FSK because it puts more information into each signal, but it requires more elaborate equipment for signal generation and detection. The PSK signal is represented as:

$$S(t) = A \cos(2\pi f_c t + \pi) \quad \text{binary 1}$$
$$S(t) = A \cos(2\pi f_c t) \quad \quad \text{binary 0}$$

PSK is used to provide multilevel modulation. The technique is called *quadrature signal modulation* (QAM). For example, a dibit modem (2 bits per baud) typically encodes the binary data stream as follows (see Figure 2.13):

11:=45°
10:=135°
01:=225°
00:=315°

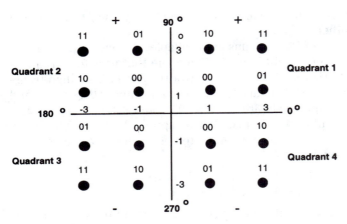

Figure 2.13 QAM.

Quadrature amplitude modulation

A special extension of multiphase PSK modulation is QAM (see Figure 2.13). The dots in Figure 2.13 represent composite signal points, and the lattice markings represent amplitude levels in each quadrature channel.

The signal points represent various combinations of amplitude and phase modulation. As Figure 2.13 shows, the points represent a phase angle and a certain level of positive or negative amplitude. The possible combinations in this simple example are any combination of 2 bits. Thus, this type of modem uses *dibits*. Alternately, the modem can use 4 bits (*quadbits*), and employ QAM techniques. Figure 2.13 also shows the four quadrants of the signals for the QAM alternative. Assume that the modem operates at 2400 bit/s at 1200 baud. The first two bits of the signal are encoded based on the preceding signal element and its position in one of the four quadrants, and the last two bits define one of the four possible bit combinations (signal elements) with the new quadrant.

The QAM and PSK spectrum shapes are identical. For example, a 16-PSK spectrum shape is the same as a 16-QAM spectrum shape. However, a QAM system exhibits considerably better error performance than its PSK counterpart. The distance between points is smaller in a PSK system. Moreover, the following expression for QAM distances between adjacent points shows that an *n*-ary QAM system performs better than an *n*-ary PSK system:

$$I = 2/(L - 1)$$

where L is the power levels on each axis.

Signal space diagrams (constellation patterns)

The depiction in Figure 2.13 is known in the ITU-T Recommendations as a *signal space diagram* (and also as a *constellation pattern* by other parties). The dots or coordinates of the signal space diagram can be labeled with binary numbers, as I have done in this figure. Each coordinate represents a set of bits, and the modem creates the proper signal to represent these bits as they are received from the user device. The transmitting modem accepts a specific number of bits (3, 4, etc.) and creates an in-phase cosine carrier and a sine wave, which are summed together and used to modulate the modem carrier signal.

Bandwidth considerations

An important factor in the analysis of digital-to-analog modulation is bandwidth efficiency—the efficiency of the bandwidth used for modulating data onto analog signals. It is expressed in the ratio of the data rate to the bandwidth. For ASK and PSK, the following formula applies:

$$B_T = (1 + r)R$$

where B_T is the transmission bandwidth, r is the filtering technique to establish the bandwidth (usually $0 < r < 1$), and R is the bit rate.

For FSK, the bandwidth is:

$$B_T = 2\Delta F + (1 + r)R$$

where $\Delta F = f_2 - f_c = f_c - f_1$, which is the offset of the modulated frequency from the carrier frequency. It can be seen that ΔF is quite significant with high-carrier frequencies.

For multilevel signalling, the bandwidth is:

$$B_T = 1 + r/\log_2 LR$$

where 1 is the number of bits that are encoded per signalling element and L is the number of signalling elements.

Connectors

In the past, the mechanical connector between the user terminal and a mode usually was a 25-position plug-type connector, published by the ISO as ISO 2110. The interchange circuits are clamped to a pin, and these pins plug into the connector. The female connector is associated with the DCE. The interface cable with the male connector is provided with the DTE. The DCE has a male shell (receptacle connector).

The combinations used on these connectors vary widely throughout the industry. Even though standards are published that state how these "pins" are to be used, many vendors ignore the standards and use their own approach. Nonetheless, the industry is in wide agreement on the physical dimensions (the mechanical aspect) of these connectors, which greatly facilitates the use of different vendors' connectors, cables, and plugs across different machines.

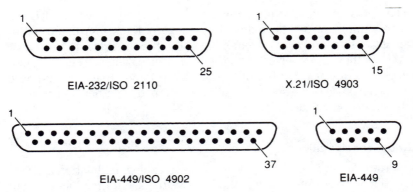

Figure 2.14 Commonly used connectors.

The widely-used EIA-232-E specification actually uses the ISO connector 2110. This connector consists of 25 pins as shown in Figure 2.14.

The EIA-449 connector is a 37-interchange circuit device. This connector is used on a limited number of DTEs because 37 interchange circuits are not needed for most physical connections.

In many installations, the EIA-232-E 25-pin connector also is impractical. Consequently, many vendors manufacture a 9-pin connector (which is described in EIA-449) and install this interface on their devices. Common examples are portable personal computers in which the space is limited on the back of the device.

The ISO-4903 connector also is a prevalent interface. The ITU-T X.21 specification utilizes this connector. However, the X.21 specification is seldom used in North America.

The ordinary telephone jack is replacing these connectors in many systems when the modem is housed inside the cabinet of the DTE. The most common jacks used for this purpose are the RJ-11, RJ-12, and RJ-13.

Other Modems

A wide variety of modems now are available in the marketplace. Most of them use the V Series specifications for their data communications interfaces. The interfaces and specifications in this section are explained in later chapters.

Fax modem

The so-called fax modem is small in size and designed for portable use and can fit in a briefcase or even a large pocket. It connects to a computer with conventional V.24, 232, or RJ-11 connectors. Typically, it uses ITU-T Group

III fax: 300 to 9600 bit/s and operates with V.21, V.22, V.22 bis, Bell 103, Bell 212A and with V.42 LAPM, V.42 bis, or MNP 5.

Telephone coupler (phone coupler)

Yet another device is the telephone or phone coupler, which was introduced earlier in this chapter. It uses audible tones and connects to a telephone handset. It is useful in hotels, etc. where phones are hard-wired and the telephone RJ connector cannot be removed. Typically, it uses the following modems: Bell 103 at 100 bit/s, Bell 212 at 1200 bit/s, V.22 at 1200 bit/s, or V.22 bis at 2400 bit/s.

Cellular modem

The cellular modem adapters are designed to operate with existing mobile cellular phones and are used at remote sites, emergency locations, etc. They support data or fax and operate from 300 to 9600 bit/s. Typically, they also support V.42 LAPM, V.42 bis, or MNP 5.

Summary

This chapter has provided an overview of the principal terms and concepts that are included in the ITU-T V Series Recommendations. You learned that the V Series machines are designed to operate on telephone-type media that exhibit analog transmission characteristics. These machines use a variety of modulation techniques and the more recent recommendations support QAM.

3

General V Series Recommendations

As the term *general* implies, these V Series recommendations provide the basic specifications and definitions that are applicable to the other V Series Recommendations. They provide definitions for symbol notations, signalling rates (in bit/s), power levels on the telephone network, etc.

This chapter is rather short, which reflects the brevity of this part of the V Series Recommendations. Figure 3.1 provides a view of the arrangement of these recommendations.

Structure of the General Recommendations

The general interface ITU-T numbers and titles are as follows:

- V.1: Equivalence between binary notation symbols and the significant conditions of a two-condition code
- V.2: Power levels for data transmission over telephone lines
- V.4: General structure of international alphabet No. 5 (IA5) code for character oriented data transmission over public telephone networks
- V.5: Standardization of data signalling rates for synchronous data transmission in the general switched telephone network (NLIF)
- V.6: Standardization of data signalling rates for synchronous data transmission on leased telephone-type circuits (NLIF)

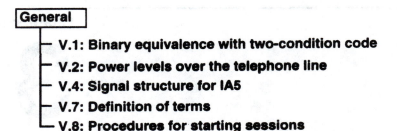

Figure 3.1   General recommendations from the V Series.

- V.7: Definitions of terms concerning data communications over the telephone network
- V.8: Procedures for starting sessions of data transmission over the switched telephone network (UCFA)

V.1

This recommendation defines the encoding and notation for binary data. Binary 0s and 1s are represented as shown in Table 3.1.

The terms Condition A and Condition Z have been retained in Recommendation V.1 to keep the signaling of 0s and 1s consistent with the telegraph service. The Condition Z in the telegraph system refers to an idle condition on the line, when signals are not being sent. Therefore, Condition Z is in the column related to the stop signal. Condition Z often is used on the older perforated tape systems where the paper tape contains a hole or perforation.

The entry dealing with phase signals simply refers to the symbol 1 being represented by the carrier signal, which experiences a phase inversion from the previous signal element. The symbol 0 is coded as a no phase inversion from the previous signal element.

V.2

This recommendation establishes the maximum power output for a telephone subscriber's equipment. It is not to exceed 1 mW at any frequency. It also specifies the maximum signal power in dBm for switched and leased lines. In addition to a description of these power levels, the reader might want to refer to the ITU-T Recommendations that describe the power output for international leased circuits.

V.4

This recommendation defines international alphabet no. 5 (IA5) as the 7-unit (bit) alphabet for use in data transmission [if the requirements cannot be met with the 5-bit international telegraph alphabet no. 2 (ITA2)]. It further establishes the use of start/stop characters for asynchronous data streams and defines the rules for the use of parity bits.

The rules for the coding of asynchronous data streams require that the character transmitted be preceded by a start signal. The start signal is considered to be a one-unit interval, which is dependent on the modulation rate of the specific V Series modem. As you might expect, the 7-bit code is followed by a parity bit and a mark signal. It also permits the use of a 2-bit stop signal for older electromechanical DTEs operating at up to 200 bauds.

Regarding parity rules, V.4 recommends that even parity be employed on lines using start/stop operations or odd parity using character-oriented synchronous formats.

V.5

V.5 establishes the signalling rates for synchronous transmission on the switched telephone network. The signalling rates are defined to be 600, 1200, 2400, 4800, and 9600 bit/s.

V.5 also stipulates the following modems for use on the switched telephone network: V.23, V.26 bis, and V.27 ter for half-duplex operations and V.22, V.22 bis, V.26 ter, and V.32 for duplex operations.

Interestingly, Recommendation V.5 no longer includes V.21 for use on the switch telephone network, but as you will see in later chapters, V.21 still is included in the Blue Books. ITU-T probably sees no point in recommending this rather antiquated 300-bit/s modem. This Recommendation is classified by the ITU-T as no longer in force (NLIF).

V.6

This recommendation, which is no longer in force (NLIF), describes the data signalling rates for synchronous transmission on leased telephone circuits. The preferred data signalling rates are 600, 1200, 2400, 4800, 9600, and 14400 bit/s. Supplementary data rates are 3000, 6000, 7200, and 12000 bit/s.

TABLE 3.1 The Representation of Binary 1s and 0s

0 Condition A	1 Condition Z
Start	Stop
Space	Mark
High frequency blank on paper tape	Low frequency hole in paper tape
Opposite phase to reference phase	Reference phase

V.6 recommends the following modems be employed on the leased telephone-type circuits:

V.22
V.22 bis
V.23
V.26
V.26 ter
V.27
V.27 bis
V.29
V.32
V.33

V.7

V.7 is a brief recommendation containing 14 definitions. One wonders why the ITU-T has not expanded the V.7 Recommendation. It provides useful information, but it is not complete when compared to the full document. (Perhaps they anticipated a book of this sort to be published to explain the ITU-T terms used in the V Series book.)

V.8

Recommendation V.8 is under consideration for approval, but as of this writing, it has not been approved by the ITU-T. It will define a new procedure for initiating a session for data transfer through a telephone connection. The intent of V.8 is to allow an answering device to pass an incoming call to another device. In addition, V.8 will include a "call menu" to help the calling and answering devices to agree on a V Series modulation mode.

Summary

This chapter has provided a summary of the ITU-T V Series General Recommendations. These recommendations provide some useful definitions and concepts for the ITU-T V Series user. The V.5 and V.6 Recommendations are cited in a number of the other V Series Recommendations. The V.2 Recommendation is used in many systems for definition of power levels on telephone lines. Other than these recommendations, the general recommendations do not play a large role in a manufacturer's equipment.

Interfaces for the V Series Recommendations and Others

This chapter examines the V Series interfaces. Several of the interfaces that are described in this chapter owe their origin to work done by a number of telecommunications carriers. However, the ITU-T provided a cohesive forum for organizing the deliberations and promulgating the results of the deliberations.

The majority of the interface recommendations include specifications for signalling between the data terminal equipment (DTE) and data circuit-terminating equipment (DCE). To a more limited extent, several of the recommendations include procedures for signalling between two DCEs.

In addition to a discussion of the V Series Recommendations, this chapter also includes comparisons between the V Series and their counterparts in the Electronic Industries Association (EIA) standards.

This part of the ITU-T V Series Recommendation is very large; it contains the Recommendations on interfaces and voice-band modems. For purposes of discussion and organization, I have divided this topic into two chapters, and chapter 5 describes the voice-band modems.

I recognize that a chapter heading containing "and Others" might seem ambiguous. My approach in dividing the ITU-T "Interfaces and voice-band modems" section into two chapters (this chapter and chapter 5) results in the fact that some of these Recommendations do not fit exactly into one of the chapters. Therefore, in this chapter, I have placed interfaces and the modems that might not be readily identifiable to the general reader (for example, medical analogue modems). The next chapter (chapter 5) com-

```
┌─────────────────────────┐
│ Interfaces and Other    │
└─────────────────────────┘
      ├─ V.10: Unbalanced interchange circuits
      ├─ V.11: Balanced interchange circuits
      ├─ V.13: Simulated carrier control
      ├─ V.14: Start-stop over synchronous channels
      ├─ V.15: Acoustic coupling
      ├─ V.16: Medical analogue modems
      ├─ V.19: Parallel transmission
      ├─ V.24: Interchange circuit definitions
      ├─ V.25: Automatic answering equipment
      ├─ V.25 bis: V.25 with 100 series
      ├─ V.28: Unbalanced interchange circuits
      ├─ V.31: Single current interchange circuits
      └─ V.31 bis: V.31 with optocouplers
```

Figure 4.1 The interface recommendations.

pletes the description of the ITU-T "Interfaces and voice-band modems" section with explanations of the "conventional" voice-band modems/services that are found in most installations.

Structure of the V Series Interfaces

The recommendations defined for the V Series interfaces are shown in Figure 4.1. From a general review of this figure, it is obvious that the majority of these recommendations define electrical signalling, operations of the interchange circuits, and procedures for automatic dial-and-answer operations.

The interface (and "others") ITU-T numbers and titles contained in Section 2 are as follows:

- V.10: Electrical characteristics for unbalanced double-current interchange circuits operating at data signalling rates nominally up to 100 kbit/s

- V.11: Electrical characteristics for balanced double-current interchange circuits operating at data signalling rates up to 10 Mbit/s

- V.13: Simulated carrier control

- V.14: Transmission of start-stop characters over synchronous bearer channels

- V.15: Use of acoustic coupling for data transmission

- V.16: Medical analogue data transmission modems

- V.19: Modems for parallel data transmission using telephone signalling frequencies

- V.20: Parallel data transmission modems standardized for universal use in the general switched telephone network (NLIF)

- V.24: List of definitions for interchange circuits between data terminal equipment and data circuit-terminating equipment

- V.25: Automatic answering equipment and/or parallel automatic calling equipment on the general switched telephone network, including procedures for disabling of echo control devices for both manually and automatically established calls

- V.25 bis: Automatic calling and/or answering equipment on the general switched telephone network (GSTN) using the 100-series interchange circuits

- V.28: Electrical characteristics for unbalanced double-current interchange circuits

- V.31: Electrical characteristics for single-current interchange circuits controlled by contact closure

- V.31 bis: Electrical characteristics for single-current interchange circuits using optocouplers

V.10

This recommendation defines interchange circuits for unbalanced systems. *Unbalanced*, in this context, means that each interface circuit has one conductor. These conductors share a common ground, and the voltage signals are measured in relation to the polarity of the voltage with respect to ground.

V.10 is intended for conventional wire interfaces between the DTE and the DCE; however, with modifications, it also is appropriate for coaxial cable. V.10 is similar to V.28 (which is discussed later in this chapter). The principal difference relates to the bit transfer rate across the interface. V.10 permits a data rate of up to 100 kbit/s.

Also, the voltage thresholds for V.10 are between +0.3 and −0.3 V. The ITU-T recommends that a V.10 cable not exceed 10 meters. However, users have extended this distance with satisfactory results. The operating distance depends on other factors, such as the quality of the media, the amount of interference surrounding the media, etc.

Figure 4.2 is a functional representation of the V.10 unbalanced interchange circuit. You should be aware that V.10 contains a number of other diagrams that allow several options beyond those that are depicted in this

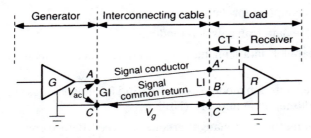

A', B' = load interchange
A, C = generator interface
   C = generator common return point
   C' = receiver zero reference point
  CT = cable termination
  GI = generator interface
  LI = load interface
V_{ac} = generator output voltage
V_g = ground potential difference

Figure 4.2 V.10. Unbalanced interchange circuit.

figure. As one example, V.10 provides for a choice in the generator (V.10 or V.11) and two categories for the receiver. The receiver can be classified as category 1, in which the receivers will have terminals A' and B' connectors of the individual terminals at the load interchange point. Category 2 defines the receivers with one terminal connection for each A' input terminal at the exchange point. In addition, all B' input terminals shall be connected together and shall be brought to one common B' input terminal.

V.10 also provides guidance on how to connect V.10 and V.28 interfaces together. You should study Annex A of V.10 for this procedure. In addition, Annex B defines the use of V.10 on coaxial cable.

V.11

V.11 specifies a balanced connection. In this context, a balanced circuit has two conductors: a signal and a return. The measurement of the voltage pertains to the difference between the signals on these two conductors.

The two V.11 wires that are associated with each interchange circuit carry signals as follows. First, one wire carries the original true form of the signal, while the other wire carries an inverted coding of the signal. This technique is known as *differential signalling* and works quite well on noisy lines because the original signal can be recovered by taking the mathematical difference between the received original signal and the received inverted signal.

V.11 interfaces usually are implemented with twisted pair cable. The wires carry the A and B forms of the signal that was just discussed. This is a useful technique because the electrical noise that is coupled into the wire is coupled equally into both wires on the interchange circuit. As just mentioned, the logic at the receiver allows the noise to be canceled out on each wire.

As with V.10, the V.11 receiver thresholds are +0.3 and –0.3 V. Remember that the difference in V.11 is that the received voltage has a difference between signal A and B, whereas with V.10 it represents the difference relative to ground.

The electrical characteristics of the V.11 allow a bit transfer rate of up to 10 Mbit/s. Consequently, this recommendation is implemented in a large number of systems that require high data transfer rate. For example, a wide variety of local area network (LAN) interfaces use the V.11 specification.

Figure 4.3 shows a functional diagram of a V.11 balanced interchange circuit.

V.13

Before I begin a discussion of V.13, you should be aware that it is necessary to explain this recommendation in the context of the V.24 interchange circuits. These interchange circuits have not yet been introduced. I have taken this approach to maintain continuity in the analysis of the recommendations. However, if you want to follow the V.13 explanation and do not know about V.24, a quick review of V.24 (which appears later in this chapter) is in order. Alternately, you might want to skip the examination of V.13 until V.24 is covered.

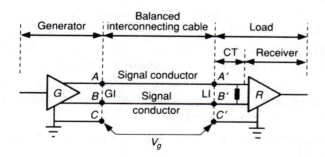

A, B, A', B' = interchange points
C, C' = 0-V reference interchange points
GI = generator interface
LI = load interface
■ = RT: optional cable termination
V_g = ground potential difference

Figure 4.3 V.11. Balanced interchange circuit.

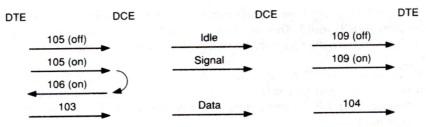

Figure 4.4 V.13 operations.

In today's modern data communications systems, a variety of full-duplex (FDX) communications systems are connected to half-duplex (HDX) devices. A good example of this configuration is the use of HDX workstations on FDX DCEs, such as multiplexers and modems.

It often is impractical to operate the FDX systems with HDX schemes. For example, the synchronization time (training time) between DCEs can be excessive if the signals must be synchronized each time that the carriers are turned off and on for the two-way alternate operation. As another example, some DCEs have no control channels between them for coordinating the control signals. To satisfy these operations, the ITU-T has published V.13.

The purpose of this recommendation is to allow a local DTE to control certain connectors on the V.24 (or EIA 232) interchange circuits of the remote DCE. In effect, a local circuit 105 request-to-send (EIA CA) controls a remote DCE interchange circuit 109 receive-line signal detect (EIA CF).

The following rules apply to the use of V.13 (see also Figure 4.4):

- When the local 105 interchange circuit (EIA CA) is off, the local DCE must transmit *idle patterns* on the link to the remote DCE. These idle patterns consist of scrambled 1s.

- When the local DTE turns on interchange circuit 105 (EIA CA), the DCE then must transmit *on patterns* to the remote DCE. The on patterns are scrambled 0s.

- At the receiving end, the remote DCE must turn off interchange circuit 109 (EIA CF) when it receives idle patterns.

- Upon detecting on patterns, the remote DCE must turn on interchange circuit 109 (EIA CF).

By adhering to these conventions, the HDX DTEs can continue to operate with FDX carrier systems.

V.14

With the advent of the personal computer, asynchronous systems have been given fresh life. Yet, many organizations use synchronous transmission techniques on the communications line. Obviously this presents problems if one must use the synchronous channel with asynchronous personal computers and workstations.

The purpose of this recommendation is to define procedures for transmitting asynchronous traffic over synchronous channels. The recommendation limits the speed over the synchronous (bearer) circuits to speeds no greater than 19.2 kbit/s.

The principal rules for converting asynchronous traffic to synchronous traffic are as follows:

- *Rule 1.* The receiving DCE receives the asynchronous characters, bounded by start/stop bits, and deletes the stop signals if the transmitting DTE is of higher speed than the DCE channel.

- *Rule 2.* Conversely, if the transmitting DTE is slower than the DCE channel, the transmitting DCE must insert additional stop signals in the transmission stream.

- *Rule 3.* The receiving DCE adjusts to the incoming bit stream by deleting or adding stop signals as appropriate to conform to the speed of its attached DTE.

Figure 4.5 shows the operations of V.14. Be aware that this figure shows asynchronous-to-synchronous-to-asynchronous conversion across the communications channel between the two DTEs. V.14 assumes that synchronous conversion is performed at the receiver. However, in practice, if the remote connection is synchronous, nothing precludes the signal remaining synchronous at that side. However, V.14 does not define this operation.

The ITU-T also has established that the V.14 Recommendations shall replace the earlier start/stop conversion to synchronous signals that were described in Recommendations V.22, V.22 bis, V.26 ter, and V.32.

V.15

V.15 describes the use of acoustic couplers for telephone lines. Acoustic couplers were quite prevalent in the 1970s but are not used much today. An acoustic coupler is a modem that sends and receives audio signals. The familiar rubber cups in a terminal is an indication that the terminal is using an

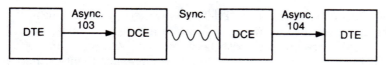

Figure 4.5 V.14 operations.

acoustic coupler modem. The ITU-T V.15 specification provides limits on the acoustic coupler's use of the telephone line and requires that the coupler be designed so that it will interwork with a standard V Series modem at the other end of the line.

The standard constricts the power output of the equipment not to exceed 1 mW at any frequency. In addition, the signal power is restricted within the voice frequency band of 0 to 4 kHz.

V.16

V.16 describes the use of computer-aided electrocardiogram transmissions on the telephone line. V.19 specifies the frequencies and power levels for the modem, acoustic coupler, and ECG. Because these instruments can operate with ECG data and other biological data on one channel, V.16 specifies subcarrier frequencies to handle multiple signalling.

V.19

V.19 is a recommendation that defines the use of parallel data transmissions. Under the V.19 Recommendation, modems can use multiple telephone frequencies to transmit digits. The transmission groups are as follows:

- Low-group frequencies: 697, 770, 852, and 941 Hz
- High-group frequencies: 1209, 1336, 1477, and 1633 Hz

As an example, the number 8 is coded by the use of two frequencies: 1336 and 852 Hz.

V.20

V.20 defines the arrangements for parallel data transmission modems for the telephone network. The frequency allocations are organized for channel numbers and for group numbers. The allocation scheme is based on 12 frequencies, consisting of four channels and three groups. This Recommendation is no longer in force (NLIF).

V.24

V.24 is one of the most widely used ITU-T V Series Recommendations in existence. Its importance stems from the fact that it defines the functions of the interchange circuits that exist between the DTE and the DCE. Obviously, without such definitions, it would be impossible to interconnect different vendors' equipment.

V.24 defines the interchange circuits for the transfer of data, as well as for control and timing signals on other interchange circuits. It specifies the operations at the DTE and the DCE. V.24 does not specify any electrical characteristics nor does it specify the physical dimensions of the interchange circuits and their connectors. These recommendations are left to other ITU-T Recommendations such as V.10 and V.11. As for the connectors, ITU-T defers to some of the ISO standards, which are discussed later in this chapter.

The V Series modems use V.24, as do other standards, such as EIA-232. In a sense, it is a "superset" standard. Vendors select the appropriate V.24 circuits for their product, as do the standards groups that publish the V Series interfaces.

V.24 is applicable to the following:

- Synchronous and asynchronous communications
- Leased or switched lines
- Two- or four-wire circuits
- Point-to-point or multipoint operation
- Certain public data networks

V.24 describes the 100 and 200 interchange circuits. The 100 series are shown in Table 4.1, and the 200 series are shown in Table 4.2. The 200 series is used for automatic calling (see V.25 and V.25 bis). The various values for telephone dialing are established in conformance to the rules in Table 4.3. The functions of the 100 and 200 series circuits are explained in the following sections.

The columns labeled "Data," "Control," and "Timing" in these tables describe the functions of the interchange circuits. Each circuit also is noted as to whether the DCE transmits ("from DCE") or receives ("to DCE") the signal.

The V.24 100 series

This section provides a brief and general description of each of the V.24 100 series interchange circuits. The reader should study V.24 for the specific rules for the use of these circuits.

Circuit 102—Signal ground or common return: This conductor establishes the common return for unbalanced circuits according to Recommendation V.28 and the dc reference for circuits according to Recommendations V.10, V.11, and V.35.

Circuit 102a—DTE common return: This conductor is used as the reference potential for the unbalanced V.10-type circuit receivers within the DCE. It is connected to the DTE common return circuit.

TABLE 4.1 V.24 100 Series Interchange Circuits

Number	Name	Ground	Data From DCE	Data To DCE	Control From DCE	Control To DCE	Timing From DCE	Timing To DCE
102	Signal Ground or Common Return	X						
102a	DTE Common Return	X						
102b	DCE Common Return	X						
102c	Common Return	X						
103	Transmitted Data			X				
104	Received Data		X					
105	Request to Send					X		
106	Ready for Sending				X			
107	Data Set Ready				X			
108/1	Connect Data Set To Line					X		
108/2	Data Terminal Ready					X		
109	Data Channel Received Line Signal Detector				X			
110	Data Signal Quality Detector				X			
111	Data Signal Rate Selector (DTE)					X		
112	Data Signal Rate Selector (DCE)				X			
113	Transmitter Signal Element Timing (DTE)							X
114	Transmitter Signal Element Timing (DCE)						X	
115	Receiver Signal Element Timing (DCE)						X	
116/1	Back up, Switching (Direct Mode)					X		
116/2	Back up, Switching (Standby Mode)					X		
117	Standby Indicator				X			
118	Transmitted Backward Channel Data			X				
119	Received Backward Channel Data		X					
120	Transmit Backward Channel Line Signal					X		
121	Backward Channel Ready				X			
122	Backward Channel Received Line Signal Detector				X			
123	Backward Channel Signal Quality Detector				X			
124	Select Frequency Groups					X		

Number	Name	Ground	Data		Control		Timing	
			From DCE	To DCE	From DCE	To DCE	From DCE	To DCE
125	Calling Indicator				X			
126	Select Transmit Frequency					X		
127	Select Receive Frequency					X		
128	Receiver Signal Element Timing (DTE)							X
129	Request-to-Receive					X		
130	Transmit Backward Tone					X		
131	Received Character Timing						X	
132	Return to Nondata Mode					X		
133	Ready for Receiving					X		
134	Received Data Present				X			
136	New Signal					X		
140	Loopback/Maintenance Test					X		
141	Local Loopback					X		
142	Test Indicator				X			
191	Transmitted Voice Answer					X		
192	Received Voice Answer				X			

TABLE 4.2 V.24 200 Series for Automatic Calling

Interchange circuit no.	Interchange circuit name	From DCE	To DCE
201	Signal Ground or Common Return	X	X
202	Call Request		X
203	Data Line Occupied	X	
204	Distant Station Connected	X	
205	Abandon Call	X	
206	Digit Signal 2^0		X
207	Digit Signal 2^1		X
208	Digit Signal 2^2		X
209	Digit Signal 2^3		X
210	Present Next Digit	X	
211	Digit Present		X
213	Power Indication	X	

TABLE 4.3 V.24 Digit Signal Circuits

Information	Binary states			
	209	208	207	206
Digit 1	0	0	0	1
Digit 2	0	0	1	0
Digit 3	0	0	1	1
Digit 4	0	1	0	0
Digit 5	0	1	0	1
Digit 6	0	1	1	0
Digit 7	0	1	1	1
Digit 8	1	0	0	0
Digit 9	1	0	0	1
Digit 0	0	0	0	0
Control Character EON*	1	1	0	0
Control Character SEP†	1	1	0	1

*The control character EON (end of number) causes the DCE to take appropriate action to await an answer from the called data station.

†The control character SEP (separation) indicates the need for a pause between successive digits or in front of the digit series, and it causes the automatic calling equipment to insert the appropriate time interval.

Circuit 102b—DCE common return: This conductor is used as the reference potential for the unbalanced V.10-type circuit receivers within the DTE. It is connected to the DCE common return circuit.

Circuit 102c—Common return: This conductor establishes the signal common return for single-current circuits with electrical characteristics according to V.31.

Circuit 103—Transmitted data: The data signals originated by the DTE, for transmittal to one or more stations, are transferred on this circuit to the DCE.

Circuit 104—Received data: The data signals generated by the DCE are transferred on this circuit to the DTE in response to data channel signals received from a data station or in response to the DTE maintenance test signals.

Circuit 105—Request to send: This circuit signals control of the data channel transmit function of the DCE. The on condition causes the DCE to assume the data channel transmit mode. The off condition causes the DCE to assume the data channel nontransmit mode when all data transferred on circuit 103 have been transmitted.

Circuit 106—Ready for sending: The signals of this circuit indicate whether or not the DCE is prepared to accept signals for transmission on the channel; the signals also might indicate that maintenance tests are under the control of the DTE. The on condition indicates the DCE is prepared to accept data signals from the DTE. The off condition indicates that the DCE is not prepared to accept data signals from the DTE.

Circuit 107—Data set ready: The signals of this circuit indicate whether or not the DCE is ready to operate. The on condition (where circuit 142 is off or is not implemented) indicates that the equipment is connected to the line and the DCE is ready to exchange data signals with the DTE for test purposes. The off condition indicates that the DCE is not ready to operate.

Circuit 108/1—Connect data set to line: This circuit signals the control switching of the signal-conversion or similar equipment to or from the line. The on condition causes the DCE to connect the signal-conversion or similar equipment to the line. The off condition causes the DCE to remove the signal-conversion or similar equipment from the line, after the transmission of all data previously transferred on circuit 103 and/or circuit 118 has been completed.

Circuit 108/2—Data terminal ready: This circuit's signals control switching of the signal-conversion or similar equipment to or from the line. The on condition prepares the DCE to connect the signal-conversion or similar equipment to the line and maintains this connection after it has been established by supplementary means. The DTE can preset the on condition of circuit 108/2 when it is ready to transmit or receive data. The off condition causes the DCE to remove the signal-conversion or similar equipment from the line, after the transmission of all data previously transferred on circuit 103 and/or circuit 118 has been completed.

Circuit 109—Data channel received line signal detector: This circuit's signals indicate whether or not the received data channel line signal is within the appropriate limits. The on condition indicates that the received signal is within appropriate limits. The off condition indicates that the received signal is not within the appropriate limits.

Circuit 110—Data signal quality detector: These signals indicate whether or not there is a reasonable probability of an error in the data received. The on condition indicates a probability of an uncorrupted transmission. The off condition indicates a reasonable probability of an error.

Circuit 111—Data signalling rate selector (DTE source): These signals are used to select one of two data signalling rates of a dual-rate synchronous or asynchronous DCE. The on condition selects the higher rate or range of rates. The off condition selects the lower rate or range of rates.

Circuit 112—Data signalling rate selector (DCE source): This circuit's signals are used to select one of the two data signalling rates in the DTE to coincide with the data signalling rate in use in a dual-rate synchronous or asynchronous DCE. The on condition selects the higher rate. The off condition selects the lower rate.

Circuit 113—Transmitter signal element timing (DTE source): The signals of this circuit provide the DCE with signal element timing information. The condition on this circuit shall be on and off for equal periods, and the transition from on to off condition indicates the center of each signal on circuit 103.

Circuit 114—Transmitter signal element timing (DCE source): The signals of this circuit provide the DTE with signal element timing information. The condition on this circuit shall be on and off for equal periods. The DTE presents a signal on circuit 103 in which the transmissions between signal elements occur simultaneously to the transitions from the off to on condition of circuit 114.

Circuit 115—Receiver signal element timing (DCE source): This circuit's signals provide the DTE with signal element timing information. The condition of this circuit shall be on and off for equal periods. The DTE presents a signal on circuit 103 in which the transitions between signal elements occur at the time of the transitions from the off to the on condition of circuit 114.

Circuit 116/1—Backup switching in direct mode: This circuit's signals are used to select the normal or standby facilities of the DCE, such as signal converters and data channels. The on condition selects the standby mode of operation, causing the DCE to replace predetermined facilities by their reserves. The off condition causes the DCE to disconnect from the standby facilities. The off condition is maintained whenever the standby facilities are not used.

Circuit 116/2—Backup switching in authorized mode: These signals are used to control the DCE between normal and standby facilities.

Circuit 117—Standby indicator: These signals indicate whether or not the DCE is conditioned to operate in standby mode with the primary facilities replaced by standbys. The on condition indicates that the DCE can operate in its standby mode. The off condition indicates that the DCE can operate in its normal mode.

Circuit 118—Transmitted backward channel data: This circuit is equivalent to circuit 103 for backward (reverse) channel transmission.

Circuit 119—Received backward channel data: This circuit is equivalent to circuit 104 for backward (reverse) channel transmission.

Circuit 120—Transmit backward channel line signal: This circuit is equivalent to circuit 105 for the backward (reverse) transmit function of the DCE. The on condition causes the DCE to assume the backward channel transmit mode. The off condition causes the DCE to assume the backward channel nontransmit mode, after all the data transferred on circuit 118 have been transmitted to the line.

Circuit 121—Backward channel ready: This circuit is equivalent to circuit 106 for backward (reverse) channel transmission. The on condition indicates that the DCE can transmit data on the backward channel. The off condition indicates that the DCE cannot transmit data on the backward channel.

Circuit 122—Backward channel received line signal detector: This circuit is equivalent to circuit 109, except that it is used to indicate that the signal quality of the received backward channel line signal is within appropriate limits.

Circuit 123—Backward channel signal quality detector: This circuit is equivalent to circuit 110, except that it is used to indicate the signal quality of the received backward channel line signal.

Circuit 124—Select frequency groups: These signals are used to select the desired frequency groups that are available in the DCE. The on condition causes the DCE to use all of the frequency groups to represent data signals. The off condition causes the DCE to use a specified reduced number of frequency groups to represent data signals.

Circuit 125—Calling indicator: These signals indicate whether or not a calling signal is being received by the DCE. The on condition indicates that a calling signal is being received. The off condition indicates no calling signal is being received.

Circuit 126—Select transmit frequency: These signals are used to select the required transmit frequency of the DCE. The on condition selects the higher transmit frequency. The off condition selects the lower transmit frequency.

Circuit 127—Select receive frequency: These signals are used to select the required receive frequency of the DCE. The on condition selects the lower receive frequency. The off condition selects the higher receive frequency.

Circuit 128—Receiver signal element timing (DTE source): These signals provide DCE with signal element timing information. The condition of this circuit is on and off for equal periods. The DCE presents a data signal on circuit 104 in which the transitions between the signal elements occur at the time of the transition from the off to the on condition of the signal on circuit 128.

Circuit 129—Request to receive: These signals are used to control the receive function of the DCE. The on condition causes the DCE to assume the receive mode. The off condition causes the DCE to assume the nonreceive mode.

Circuit 130—Transmit backward tone: These signals control the transmission of a backward channel tone. The on condition causes the DCE to transmit a backward channel tone. The off condition causes the DCE to stop the transmission of a backward channel tone.

Circuit 131—Received character timing: These signals provide the DTE with character-timing information. The specific V Series Recommendation should be checked.

Circuit 132—Return to nondata mode: These signals are used to restore the nondata mode that is provided with the DCE without releasing the line connection to the remote station. The on condition causes the DCE to restore the nondata mode. When the nondata mode has been established, the circuit is turned off.

Circuit 133—Ready for receiving: These signals control the transfer of data on circuit 104, indicating whether the DTE is capable of accepting a given amount of data as specified in the recommendation for intermediate equipment. The on condition is maintained whenever the DTE is capable of accepting the data and causes the intermediate equipment to transfer the received data to the DTE. The off condition indicates that the DTE is not able to accept the data and causes the intermediate equipment to retain the data.

Circuit 134—Received data present: This signal is used to separate information messages from supervisory messages that are transferred on circuit 104. The on condition indicates the data that represent information messages. The off condition shall be maintained at all other times.

Circuit 136—New signal: These signals are used to control the response times of the DCE receiver. The on condition instructs the DCE receiver to prepare itself to rapidly detect the disappearance of the line signal (by disabling the response time circuitry that is associated with circuit 109). After the received line signal falls below the threshold of the received line signal detector, the DCE will turn off circuit 109 and prepare itself to rapidly detect the appearance of a new line signal.

Circuit 140—Loopback/maintenance test: These signals are used to initiate and release loopback or other maintenance test conditions in DCEs. The on condition causes initiation of the maintenance test condition. The off condition causes release of the maintenance test condition.

Circuit 141—Local loopback: These signals are used to control the loop 3 test condition in the local DCE. The on condition of circuit 141 causes the establishment of the loop 3 test condition in the local DCE. The off condition of circuit 141 causes the release of the loop 3 test condition of the local DCE.

Circuit 142—Test indicator: These signals indicate whether or not a maintenance condition exists. The on condition indicates that a maintenance condition exists in the DCE, precluding reception of transmission of data signals from or to a remote DTE. The off condition indicates that the DCE is not in a maintenance test condition.

Circuit 191—Transmitted voice answer: These signals, which are generated by a voice answer unit in the DTE, are transferred on this circuit to the DCE.

Circuit 192—Received voice answer: The received voice signals, which are generated by a voice answering unit at the remote DTE, are transferred on this circuit to the DTE.

Examples of the use of the V.24 interchange circuits

Figures 4.6 and 4.7 provide examples of the use of the V.24 interchange circuits and the relationship to the EIA-232-E standard. The arrows symbolize

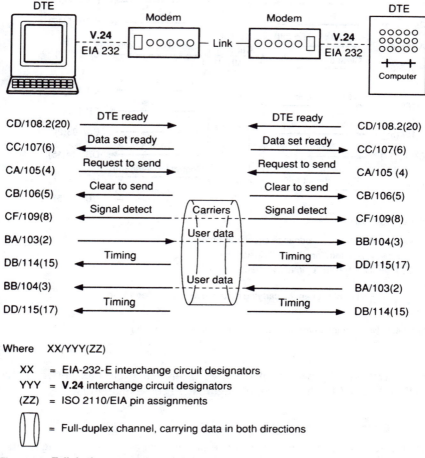

Where   XX/YYY(ZZ)

XX = EIA-232-E interchange circuit designators
YYY = **V.24** interchange circuit designators
(ZZ) = ISO 2110/EIA pin assignments

= Full-duplex channel, carrying data in both directions

Figure 4.6 Full-duplex operations with V.24 interchange circuits.

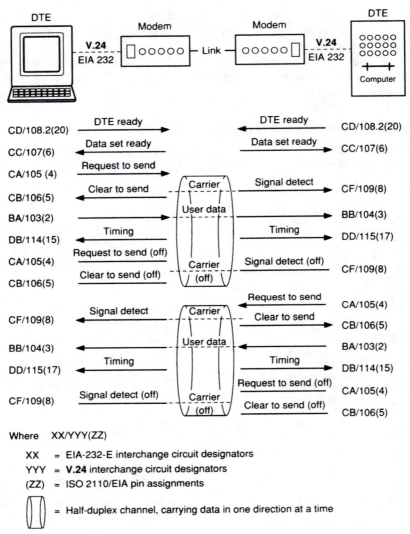

Where   XX/YYY(ZZ)

XX   = EIA-232-E interchange circuit designators
YYY   = V.24 interchange circuit designators
(ZZ) = ISO 2110/EIA pin assignments

      = Half-duplex channel, carrying data in one direction at a time

Figure 4.7   Half-duplex operations with V.24 interchange circuits.

the activation/deactivation of the V.24 interchange circuits between the DTE and the DCE. Otherwise, the figures are self-explanatory.

V.24 200 series

As stated earlier, the 200 series is used to control automatic calling operations. Tables 4.2 and 4.3 provide information on these circuits.

Circuit 201—Signal ground or common return: This conductor establishes the signal common reference potential for all 200-series interchange circuits.

Circuit 202—Call request: These signals are used to condition the automatic calling equipment to originate a call and to switch the automatic calling equipment to or from the line. The on condition causes the DCE to condition the automatic calling equipment to originate a call and to connect this equipment to the line. The off condition causes the automatic calling equipment to be removed from the line and indicates that the DTE has released the automatic calling equipment.

Circuit 203—Data line occupied: These signals indicate whether or not the associated line is in use (e.g., for automatic calling, data transmission, voice communication, or test procedures). The on condition indicates that the line is in use. The off condition indicates that the line is not in use and that the DTE can originate a call.

Circuit 204—Distant station connected: These signals indicate whether or not a connection has been established to a remote station. The on condition indicates the receipt of a signal from a remote DCE signalling that a connection to that equipment has been established. The off condition shall be maintained at all other times.

Circuit 205—Abandon call: These signals indicate whether or not a preset time has elapsed between successive events in the calling procedure. The on condition indicates that the call should be abandoned. The off condition indicates that the call origination can proceed.

Digital signal circuits: On circuits 206 [digital signal (2^0)], 207 [digital signal (2^1)], 208 [digital signal (2^2)], and 209 [digital signal (2^3)], the DTE presents the code combinations for the binary dialing digits.

Circuit 210—Present next digit: These signals indicate whether or not the automatic calling equipment is ready to accept the next code combination. The on condition indicates that the automatic calling equipment is ready to accept the next code combination. The off condition indicates that the automatic calling equipment is not ready to accept signals on the digit signal circuits.

Circuit 211—Digit present: These signals control the reading of the code combination that is presented on the digit signal circuits. The on condition causes the automatic calling equipment to read the code combination that is presented on the digit signal circuits. The off condition prevents the automatic calling equipment from reading a code combination on the digit signal circuits.

Circuit 213—Power indication: These signals indicate whether or not power is available within the automatic calling equipment. The on condition indicates that power is available within the automatic calling equipment. The off condition indicates that power is not available within the automatic calling equipment.

TABLE 4.4 Dial and Answer Interchange Circuits

Interchange circuit number	Description
104	Received Data
105	Request to Send
106	Ready for Sending
107	Data Set Ready
108/1	Connect Data Set to Line
108/2	Data Terminal Ready
109	Data Channel Received Line Signal Detector
119	Received Backward Channel Data
120	Transmit Backward Channel Line Signal
121	Backward Channel Ready
122	Backward Channel Received Line Signal Detector
125	Calling Indicator
201	Signal Ground or Common Return
202	Call Request
203	Data Line Occupied
204	Distant Station Connected
205	Abandon Call
206	Digit Signal (2^0)
207	Digit Signal (2^1)
208	Digit Signal (2^2)
209	Digit Signal (2^3)
210	Present Next Digit
211	Digit Present
213	Power Indication
EON	End-of-Number Control Character
SEP	Separation Control Character

V.25

V.25 describes the conventions for automatic calling and answering. The ITU-T dial-and-answer modems (chapter 5) use the V.25 specification, but vendors might vary in their use and interpretation of V.25. A careful review of their specifications is necessary to determine compatibility of equipment.

As with other V Series interfaces, the V.24 circuits are used in the V.25 specification. It also makes use of the 200 series interchange circuits of V.24. In addition, V.25 includes procedures for disabling echo control devices. The V.24 100 and 200 series circuits are defined for the V.25 functions listed in Table 4.4.

To establish, maintain, and disconnect the call, the DTE is responsible for the following tasks:

- Ensuring that the DCE is available
- Providing the telephone number

- Abandoning the call if necessary
- Establishing proper identification
- Exchanging traffic
- Initiating a disconnect

The procedures for the calling and answering stations are established by V.25 as shown in Table 4.5.

TABLE 4.5 Automatic Dial and Answer

	Calling Station Event
1	DTE checks whether CT 213 is on and the following circuits are off: CT 202, CT 210, CT 205, CT 204, and CT 203.
2	DTE turns CT 202 on.
3	DTE turns CT 108/2 on.
4	For half-duplex modems, DTE puts CT 105 on if the calling end wishes to transmit first.
5	Line goes off hook, which has the same effect as picking up your telephone.
6	DCE turns CT 203 on.
7	Telephone office or PBX puts a dial tone on the line.
8	DCE turns CT 210 on.
9	DTE presents the appropriate telephone number digit on CT 206, CT 207, CT 208, and CT 209. (Four circuits are needed to represent a base 10 number of 0 through 9 in binary base 2.)
10	DTE turns CT 211 on after digit signals have been presented.
11	DCE dials first digit, then turns CT 210 off.
12	DTE turns CT 211 off.
13	Events 8 to 12 are repeated (which can be interrupted by SEP) until the last digit signal is presented and transferred. Event 8 then is repeated, but event 14 follows.
14	DTE presents EON on CT 206, CT 207, CT 208, and CT 209; it then turns CT 211 on.
15	DCE turns CT 210 off.
16	DTE turns CT 211 off and turns CT 108/2 on, if not previously on.
17	A calling tone is transmitted from calling DCE.
18	When calling DCE recognizes that an answer has occurred, it stops the calling tone and transfers control of the line from CT 202 to CT 108/2.

	Called Station Event
1	A ring signal is detected on the line. DCE turns on CT 125.
2	If CT 108/2 is on, DCE goes off hook. However, if CT 108/1 or CT 108/2 is off, DCE waits for one of them to come on. Otherwise, the call is not answered.
3	The DCE, upon going off hook, does nothing for a period of 1.8 to 2.5 seconds.
4	DCE transmits a 2100-Hz (±15 Hz) answer tone. This tone can disable echo suppressors by performing 180° phase reversals of the tone at 425- to 475-ms intervals.
5	The 2100-Hz tone is discontinued upon the calling station responding for 100 ms. (This specification can be relaxed for "slower" devices such as acoustic modems.)
6	After 75 ms (±20 ms), the DCE turns on CT 107.

Stations operating in a manual mode are handled by a procedure that is quite similar to the automatic sequences, but no tone is transmitted from the calling station. The person who dials the number, on hearing the 2100-Hz signal from the called station, presses the "data" button to connect the DCE to the telephone line. Then CT 107 comes on.

V.25 bis

This recommendation uses the same concepts as V.25, except V.25 bis uses serial transmission. Unlike V.25, the bis version uses only one circuit (V.24 circuit 103) to present a dialed digit.

V.25 bis is a common interface for workstations and personal computers. The V.24 circuits used by V.25 bis are shown in Table 4.6.

V.25 bis provides for three types of calls:

- A serial automatic calling data station to an automatic answering data station

- A manual calling data station to an automatic answering data station

- A serial automatic calling data station to a manual answering data station

In a manner similar to the V.25 Recommendation, the V.25 bis DTE is responsible for:

- Ensuring that the DCE is available

- Providing the telephone number or selecting a telephone number stored at the DCE

- Deciding to abandon the call

- Controlling data transfer

- Initiating disconnect at calling or answering data stations

TABLE 4.6 V.25 bis Interchange Circuits for Automatic Calling and Answering

No.	Interchange circuit name	Direction	
		From DCE	To DCE
103	Transmitted Data and Commands		X
104	Received Data and Indications	X	
106	Ready for Sending	X	
107	Data Set Ready	X	
108/2	Data Terminal Ready		X
125	Calling Indicator	X	

TABLE 4.7 V.25 bis Commands and Indications

Command or indication	DTE to DCE (command)	DCE to DTE (indication)	Parameters
Call Request	X		Number to be Dialed
			Memory Address of the Number to be Dialed
			Double Dial-up Request
			Identification Number
Program	X		Number to be Dialed
			Memory Address for the Number to be Dialed
			Identification Number
List Request	X		Under study
Disregard Incoming Call	X		None
Connect Incoming Call	X		None
Call Failure		X	Engaged Tone
			Number Not Stored
			Local DCE Busy
			Ring Tone (time out)
			Abort Call (time out)
			V.25 Answer Tone Not Tested
			Forbidden Call
Delayed Call		X	Time to Permissible Call Request (minute)
Incoming Call		X	None
Valid		X	None
Invalid		X	Optionally an Error Code
List		X	Memory Address
			Number to be Dialed
			Identification Number
			Status (under study)
Call Connecting		X	Under study

V.25 bis provides for several commands and indications. In this way, the DTE instructs the DCE to enter into a specific state (operation). These commands are accompanied by instructional parameters shown in Table 4.7.

V.25 bis permits the commands and indications to be transmitted in one of three ways:

- Asynchronous
- Synchronous character oriented
- Synchronous bit oriented

V.28

V.28 specifies the electrical characteristics of the interface between the DTE and DCE. That is to say, it describes the characteristics of the receivers and drivers that are used across the V.24 interchange circuits. V.28 requires the following electrical specifications be established for the interface:

- The dc resistance in the receiver must be between 3000 and 7000 Ω.
- The receiver must not exceed 2500 pF at the interchange point.
- The driver should not exceed 25 V. It should generate a voltage at the interchange point between 5 and 15 V for a load resistance in the range of 3000 and 7000 Ω.
- Controlling timing circuits are defined with off/on states. The on state is indicated by a signal that is more positive than +3 V. The off state is indicated with a signal which is more negative than –3 V.
- For the data interchange circuits, a binary 1 is sent as more negative than –3 V. A binary 0 is represented by a more positive than +3 V.
- The region between the +3 and –3 V is a transition region. The signal state is not defined in this region.
- V.28 is limited to a signaling rate of up to 20 kbit/s.

V.31

V.31 describes a current loop interface. The current loop has been used for many years for machines such as mechanical typewriters, some asynchronous computer ports, and the older telex devices. It represents signals by using the presence of a data flow to convey a binary 0 or 1.

V.31 defines a current loop interface for data rates below 75 bit/s. It is similar to the North American standard published by the EIA and designated as RS-410.

The V Series, the ISO Connectors, and EIA-232

As explained in chapter 2, the ISO publishes many standards, some of which describe the mechanical connectors that are used by computers, terminals, modems, and other devices (see Figure 2.14). The ITU-T V Series do not stipulate a ITU-T-defined connector. Rather, they stipulate an ISO connector, typically one of the connectors shown in Figure 2.14. The EIA also uses the ISO connector standards in many of its documents.

EIA-232-E (and its predecessors EIA-232-D and RS-232-C) is one of the most widely used physical interfaces in the world. It is sponsored by the EIA and is quite prevalent in North America.

As you learned earlier in this chapter, the ITU-T publishes the V.24 specification, from which the ISO connector arrangements and EIA-232-E pin assignments are derived. The ISO establishes the specification for the mechanical dimensions of the pins and connector, and EIA-232-E uses the ISO

2110 connector. EIA-232-E also uses two of the V Series standards: V.24 and V.28.

At first glance, the EIA-232-E interface seems very confusing, but it simply uses the following ITU-T/ISO recommendations and standards:

Electrical   ITU-T V.28
Functional   ITU-T V.24
Mechanical   ISO 2110
Procedural   ITU-T V.24

EIA-232-E and the ITU-T V.24 and V.28 standards

The V Series modems use V.24, as do standards such as EIA-232-E, although EIA-232-E uses different designations for the circuits. V.28 is applied to almost all interchange circuits operating below the limit of 20 kbit/s. The recommendation provides specifications for other electrical interfaces as well. EIA-232-E uses this specification with some minor variations. On a general level, the signals must conform to the characteristics described in the following paragraphs.

For data interchange circuits, the signal is in the binary 1 condition when the voltage on the interchange circuit that is measured at the interchange point is more negative than –3 V. The signal is in the binary 0 condition when the voltage is more positive than +3 V.

For control and timing interchange circuits, the circuit is on when the voltage on the interchange circuit is more positive than +3 V and shall be considered off when the voltage is more negative than –3 V.

Table 4.8 describes the interchange circuits of EIA-232-E and their relationships to ITU-T V.24. These circuits are 25-pin connections and sockets (refer to Figure 2.14 in chapter 2). The terminal pins plug into the modem sockets. All of the circuits are rarely used; many modems use only 4 to 12 pins. You might want to review Figures 4.6 and 4.7, which show how the EIA-232-E circuits are employed (and their relationships to V.24).

The EIA-232-E circuits perform one of four functions:

- Data transfer across the interface
- Control of signals across interface
- Clocking signals to synchronize data flow and regulate the bit rate
- Electrical ground

V.31 bis

V.31 bis describes the V.24 interchange circuits that operate at rates up to 75 bit/s (which are explained in V.31). In addition, V.31 bis describes operations on interchange circuits for rates up to 1200 bit/s. The specification

TABLE 4.8 EIA-232-E Interchange Circuits

Inter-change circuit	ITU-T equiva-lent	Description	Ground	Data From DCE	Data To DCE	Control From DCE	Control To DCE	Timing From DCE	Timing To DCE
AB	102	Signal Ground or Common Return	X						
BA	103	Transmitted Data			X				
BB	104	Received Data		X					
CA	105	Request to Send					X		
CB	106	Clear to Send				X			
CC	107	DCE Ready				X			
CD	108.2	DTE Ready					X		
CE	125	Ring Indicator				X			
CF	109	Received Line Signal Detector				X			
CG	110	Signal Quality Detector				X			
CH	111	Data Signal Rate Selector (DTE)					X		
CI	112	Data Signal Rate Selector (DCE)				X			
DA	113	Transmitted Signal Element Timing (DTE)							X
DB	114	Transmitted Signal Element Timing (DCE)						X	
DD	115	Receiver Signal Element Timing (DCE)						X	
SBA	118	Secondary Transmitted Data			X				
SBB	119	Secondary Received Data		X					
SCA	120	Secondary Request to Send					X		
SCB	121	Secondary Clear to Send				X			
SCF	122	Secondary Received Line Signal Detector				X			
RL	140	Remote Loopback					X		
LL	141	Local Loopback					X		
TM	142	Test Mode				X			

consists of three pages that describe single current interchange circuits using optocouplers (i.e., two conductors with go and return leads that are electrically insulated from each other; therefore, a common return lead can be assigned to more than one interchange circuit). You can consult the V.31 bis Recommendation for more information on the limits for current, capacitance, and resistance for the interchange circuits.

Summary

The ITU-T V Series interfaces are used extensively throughout the world. They are found in practically all vendors' modems and multiplexers and in some data service units (DSUs). They define the nature of the physical connection of user devices to communication machines. The V.24 Recommendation is widely used throughout the industry. Dial-and-answer systems have migrated to V.25 or V.25 bis. Equally important are the electrical specifications, principally V.10, V.11, and V.28. This chapter also has provided a summary of the relationships of the V.24 interchange circuits with the EIA 232-E interchange circuits.

5

The Voice-Band Modems Recommendations

It is quite likely that people who use modems are familiar with the V Series Recommendations on voice-band modems. These specifications define the operations for data transmission over the voice-grade telephone network or over circuits with similar characteristics. They define the signalling between data circuit-terminating equipment (DCEs) and between data terminal equipment (DTEs) and DCEs. They are used in practically all vendors' products. Indeed, it is rare for any modem *not* to use these standards.

As you learned in chapter 4, the V Series Recommendations in Section 2 dealing with interfaces and voice-band modems are grouped together. My approach is to examine the interfaces in chapter 4, the voice-band modems in this chapter, and the wideband modems in chapter 6. Each modem standard is described, and a table is provided at the end of each description to summarize the principal operating characteristics of the modem.

The Modem Market

The modem market has changed dramatically since its inception in the early 1960s. Before discussing each of the V Series modems, it might be useful to take a brief look at some of the major changes.

A large portion of the modem market was influenced by the introduction in 1973 of a switched, full-duplex 1200-bit/s modem by Racal-Vadic Corporation. This modem made a large impact on the industry and was followed later by the Bell 212 modem.

A significant aspect of the Racal-Vadic VA3400 was a buffer arrangement that allowed variable data rates of 0 to 300 bit/s or 1200 bit/s. This operation could be accomplished without any cumbersome switching or reconfiguration. The Bell 212 included capabilities to operate with the earlier Bell 103; therefore, it was compatible with the older Bell modems. Because of this feature, the Bell products eventually displaced the VA3400.

In the late 1970s and early 1980s, many modem manufacturers introduced the ITU-T V.22 bis Recommendation into their product lines. This modem provided a higher-speed operation of 2400 bit/s. In these early days, few people clamored for a higher-rate modem because a 1200-bit/s rate was more than sufficient to handle interactive applications in which individuals entered data manually and read messages from the terminal screen. However, with the increased use of personal computers and the frequent need for file transfer between PCs and mainframes, higher speeds were needed and the V.22 bis offered an economical, flexible choice.

A major impact on the modem market has been made by the introduction of a 9.6-kbit/s full-duplex, dial-up modem that is configured around the ITU-T V.32 specifications. Later chapters also analyze the operations and

Voice-band modems

— V.17: 7200-14400 modem on GSTN 2-wire for fax
— V.18: Internetworking modems in text telephone mode
— V.21: 300 b/s duplex modem on GSTN
— V.22: 1200 b/s duplex modem on GSTN & P-P 2-wire circuits
— V.22 bis: 2400 b/s duplex modem on GSTN & P-P 2-wire circuits
— V.23: 600/1200 baud modem on GSTN
— V.26: 2400 b/s modem on 4-wire leased circuits
— V.26 bis: 1200/2400 b/s modem on GSTN
— V.26 ter: 2400 b/s modem duplex modem on GSTN & P-P 2-wire
— V.27: 4800 b/s modem on leased circuits
— V.27 bis: 2400/4800 b/s modem on leased circuits
— V.27 ter: 2400/4800 b/s modem on GSTN
— V.29: 9600 b/s modem on P-P 4-wire circuit
— V.32: 4800-9600 b/s modem on GSTN and leased circuits
— V.32 bis: 4800-14400 b/s modem on GSTN and 2-wire leased
— V.33: 14400 b/s modem on P-P 4-wire circuits
— V.34: 14400 b/s modem on GSTN or P-P 2-wire circuits

Figure 5.1 Voice-band modems.

impact of the V.42 and V.42 bis Recommendations. They have been combined with V.32 to provide for high-speed modems with powerful error correction and data compression capabilities.

The V.29 Recommendations also have made a major impact on the industry because they have been implemented with many of the facsimile machines (specifically, the ITU-T group 3 Recommendation). Consequently, V.29 has played a major role in the industry as well.

Structure of the V Series voice-band modems

The voice-band modems that are described in Section 2 are listed here (see also Figure 5.1):

- V.17: A two-wire modem for facsimile applications with rates up to 14,400 bits/s

- V.18: Operational and interworking requirements for modems operating in the text telephone mode

- V.21: 300 bit/s duplex modem standardized for use in the general switched telephone network (GSTN)

- V.22: 1200 bit/s duplex modem standardized for use in the general switched telephone network and on point-to-point, two-wire leased telephone-type circuits

- V.22 bis: 2400 bit/s duplex modem using the frequency division technique standardized for use on the general switched telephone network and on point-to-point, two-wire leased telephone-type circuits

- V.23: 600/1200 baud modem standardized for use in the general switched telephone network

- V.26: 2400 bit/s modem standardized for use on four-wire leased telephone-type circuits

- V.26 bis: 2400/1200 bit/s modem standardized for use in the general switched telephone network

- V.26 ter: 2400 bit/s duplex modem using the echo cancellation technique standardized for use on the GSTN and on point-to-point, two-wire leased telephone-type circuits

- V.27: 4800 bit/s modem with manual equalizer standardized for use on leased telephone-type circuits

- V.27 bis: 4800/2400 bit/s modem with automatic equalizer standardized for use on leased telephone-type circuits

- V.27 ter: 4800/2400 bit/s modem standardized for use in the general switched telephone network

- V.29: 9600 bit/s modem standardized for use on point-to-point, four-wire leased telephone-type circuits
- V.32: A family of two-wire, duplex modems operating at data signalling rates of up to 9600 bit/s for use on the GSTN and on leased telephone-type circuits
- V.32 bis: A duplex modem operating at data signalling rates of up to 14,400 bit/s for use on the general switched telephone network and on leased point-to-point two-wire telephone-type circuits
- V.33: 14,400 bit/s modem standardized for use on point-to-point four-wire leased telephone-type circuits
- V.34: A modem operating at data signalling rates of up to 28,800 bit/s for use on the general switched telephone network and on leased point-to-point, two-wire telephone-type circuits

The remainder of this chapter describes each V Series voice-band modem. A table is located at the end of each explanation that summarizes the major characteristics of the modem. Each table contains several abbreviations, which are explained in Table 5.1.

Trellis Coding and Signal Structures

The V.17, V.32, V.32 bis, and V.33 modems employ similar operations for their coding and signal structures. This section provides a summary and

TABLE 5.1 Explanation of Abbreviations in the V Series Modem Tables

Heading			
Nr. = series number	Speed = bit/s	Rate = modulation rate	Rat = bits per baud
Sep = channel division	Car = carrier	H/F = half/full duplex	S/A = sync./async.
Mod = modulation	B? = backward channel	Sw? = switched line	Ded? = dedicated line
Equa = equalization			
Notes: where XXXXX = use of V.2, V.28, 2110, Scrambler, V.25			

In table		
Y = yes	N = no	NA = not applicable
2W = two wire	4W = four-wire	FD = frequency division
O = optional	Ne = none	QAM = quadrature amplitude modulation
MP = multipoint	PP = point to point	TCM = trellis coded modulation
A = adaptive	F = fixed	FS = frequency shift
E = either	M = manual	D = definition not specified
Fu = further study	EC = echo cancellation	PS = phase shift

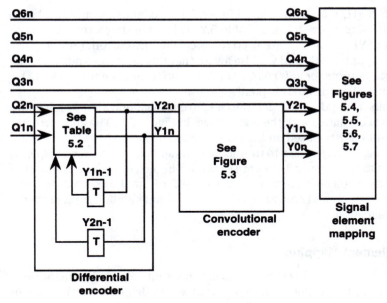

Figure 5.2 Coding architecture.

comparison of these operations. This approach is taken to avoid repeating the material in each section dealing with these modems.

To aid you in following this discussion, these modems are capable of operating at the following rates:

- V.17: 7.2, 9.6, 12, and 14.4 kbit/s
- V.32: 2.4, 4.8, and 9.6 kbit/s
- V.32 bis: 4.8, 7.2, 9.6, 12, and 14.4 kbit/s
- V.33: 14.4 kbit/s

Figure 5.2 illustrates the coding logic for these modems. All of them use this scheme. Note that the function labeled "Signal element mapping" refers to other material in this chapter that explains how the modems create the signal structure.

The user data stream is divided into bit groups labeled $Q1_n$ through $Q6_n$ in Figure 5.2, where n designates the sequence number of the bit in each group. The manner in which the bits are grouped depends upon the bit rate of the modem. As examples, a V.32 9600 bit/s modem accepts bits $Q1_n$ through $Q4_n$, which you learned earlier in this book are called *quadbits*. Because V.32 operates with a modulation rate of 2400 baud, the four bits per baud provide a 9600 user data rate. A V.32 bis modem at 14,400 bit/s accepts bits $Q1_n$ through $Q6_n$ ($6 \times 2400 = 14,400$).

Bits $Q1_n$ and $Q2_n$ are coded in a differential encoder as $Y1_n$ and $Y2_n$, based on the encoding rules shown in Table 5.2. This table shows that the previous outputs $Y1_{n-1}$ and $Y2_{n-1}$ are used to determine the current output of the coding of input bits $Q1_n$ and $Q2_n$. Be aware that this table does not show the $Y1_n$ and $Y2_n$ outputs for 4800 bit/s, which are different from this table. Later discussions will return to this point.

Next, bits $Y1_n$ and $Y2_n$ are fed into a convolutional encoder, which codes the bits in accordance with the logic shown in Figure 5.3. The coder creates a redundant bit $Y0_n$, based on the rules in Figure 5.3.

This architecture pertains to the modems under discussion. However, some exceptions to these general rules are explained in each Recommendation. For example, the 4800 bit/s $Y1_n$ and $Y2_n$ output differ from those shown in Table 5.2. The reader should study the relevant Recommendation for a more detailed explanation.

Signal Element Mapping

As Figure 5.2 shows, the Q and Y bits are input into a signal element mapping function, which generates the signal space diagram for the various modulation signals. This section provides examples of signal element mapping. They are not all inclusive but meant to show the reader several constellation patterns.

TABLE 5.2 Rules for Encoding Bits $Q1_n$ and $Q2_n$

Inputs		Previous outputs		Outputs	
$Q1_n$	$Q2_n$	$Y1_{n-1}$	$Y2_{n-1}$	$Y1_n$	$Y2_n$
0	0	0	0	0	0
0	0	0	1	0	1
0	0	1	0	1	0
0	0	1	1	1	1
0	1	0	0	0	1
0	1	0	1	0	0
0	1	1	0	1	1
0	1	1	1	1	0
1	0	0	0	1	0
1	0	0	1	1	1
1	0	1	0	0	1
1	0	1	1	0	0
1	1	0	0	1	1
1	1	0	1	1	0
1	1	1	0	0	0
1	1	1	1	0	1

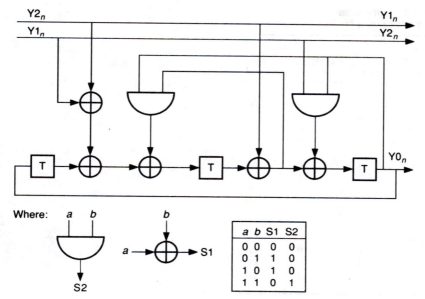

Where:

a b	S1	S2
0 0	0	0
0 1	1	0
1 0	1	0
1 1	0	1

Figure 5.3 The convolutional coder.

Mapping for 2400 bit/s

This part of the section is somewhat anticlimactic. The only modem under discussion that (supposedly) supports a 2400 bit/s rate is V.32, and the signal element coding for 2400 bit/s is for further study. It is unlikely that any more work will be done on this low data rate.

Mapping for 4800 bit/s

This rate is supported by V.32 and V.32 bis. For both modems, the scrambled data stream is grouped into the $Q1_n$ and $Q2_n$ bits, and are differentially encoded into $Y1_n$ and $Y2_n$ in accordance with Table 5.2. Figure 5.4 shows the signal space diagram for V.32; the A, B, C, and D states in Figure 5.4 can be correlated with Table 5.3.

Mapping for 7200 bit/s

At 7200 bit/s rates for V.17 and V.32 bis, the scrambled data stream is divided into groups of three bits $Q1_n$, $Q2_n$, and $Q3_n$ according to Figure 5.2. The $Y0_n$, $Y1_n$, $Y2_n$, and $Q3_n$ bits are mapped into the 16-point signalling structure shown in Figure 5.5. The binary numbers in this pattern refer to

**TABLE 5.3 Differential
Quadrant Coding for 4800 bit/s**

Phase quadrant change	Outputs $Y1_n$	$Y2_n$	Signal state
+90°	0	1	B
	1	1	C
	0	0	A
	1	0	D
0°	0	0	A
	0	1	B
	1	0	D
	1	1	C
+180°	1	1	C
	1	0	D
	0	1	B
	0	0	A
+270°	1	0	D
	0	0	A
	1	1	C
	0	1	B

these bits. The A, B, C, and D notations refer to synchronizing signal elements. These elements are used in a number of combinations for starting up the training procedure between the modems, estimating the round trip delay between the modems, training of echo cancelers (if necessary for half-duplex transmission), and selecting the bit-rate for the connection.

Mapping for 9600 bit/s

Figure 5.6 shows an example of a 9600 bit/s signal space diagram. This example is V.32 with trellis coding. As discussed earlier, the user data stream is grouped into four bits and fed into the differential coder, convolutional

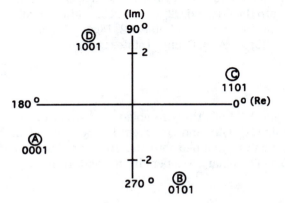

Figure 5.4 V.32 4800 bit/s signal structure.

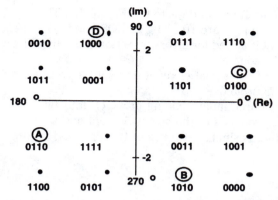

Figure 5.5 Signal space diagram and mapping for modulation at 7200 bit/s.

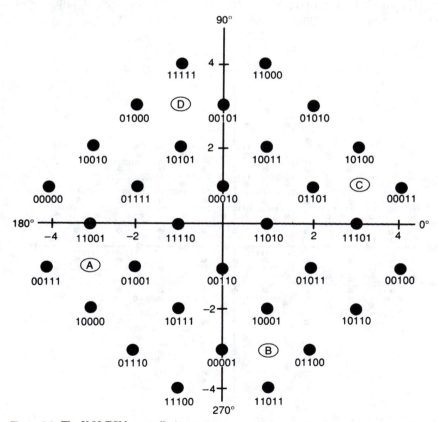

Figure 5.6 The V.32 TCM constellation pattern.

coder, and signal space mapping function depicted in Figure 5.2. The bits that are input into the space mapping are $Y0_n$, $Y1_n$, $Y2_n$, $Q3_n$, and $Q4_n$. The $Y0_n$ bit is the redundant bit that is added for error correction (see chapter 1 for more information on trellis code modulation). These five bits then are mapped into the coordinates of the signal element and sent according to the space diagram shown in Figure 5.6.

Additionally, V.32 permits nonredundant coding in which four bits (no $Y1_n$ bit) are coded into the coordinates of the signal element and sent according to a space diagram that is similar to (but not the same as) Figure 5.5.

Mapping for 12,000 bit/s

The V.32 bis modem can transmit and receive at 12 kbit/s. The technique for this rate is quite similar to V.32, except that the data stream is divided into groups of five consecutive data bits (see Figure 5.2) and that bits $Y1_n$, $Y2_n$, $Q3_n$, $Q4_n$, and $Q5_n$ are mapped into the coordinates of a signal element and sent according to a signal space diagram (not shown here).

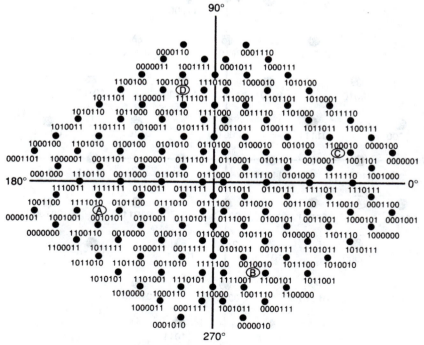

Figure 5.7 The V.33 constellation pattern.

TABLE 5.4 Principal Characteristics of V.17

Nr.	Speed	Rate	Rat	Sep	Car	H/F	S/A	Mod	B?	Sw?	Ded?	Equa	Notes
V.17	14400	2400	7	EC	1800	F	S	TCM	D	Y2W	N	Y	DDDYD
V.17	12000	2400	6	EC	1800	F	S	TCM	D	Y2W	N	Y	DDDYD
V.17	9600	2400	5	EC	1800	F	S	TCM	D	Y2W	N	Y	DDDYD
V.17	7200	2400	4	EC	1800	F	S	TCM	D	Y2W	N	Y	DDDYD

Mapping for 14,400 bit/s

An example of a 14,400 bit/s modem is provided in Figure 5.7. For this operation, all of the bits shown in Figure 5.2 are used at the signal element mapping function to produce the pattern shown in Figure 5.7.

This concludes the introduction to trellis coding and signal structures. Later discussions will revisit trellis coding. However, you now have sufficient background information to examine the V Series voice-band modems.

V.17

V.17 defines the operations for a two-wire modem for facsimile transmissions. The specification supports data rates up to 14.4 kbit/s and these other speeds: 12 kbit/s, 9.6 kbit/s, and 7.2 kbit/s. QAM is employed on synchronous transmissions using 2400 baud. V.17 includes scrambling, adaptive equalizers, and trellis coded modulation. Like other high-speed ITU-T modems, V.17 operates with an 1800-Hz carrier frequency.

V.17 employs some of the same constellation patterns and trellis coding that is used by V.32, V.32 bis, V.33, which will be explained shortly. It also uses 2400 baud and operates with a carrier frequency of 1800 Hz.

Because many of the trellis coding schemes were derived from V.32 and V.33, I previously discussed these operations in the section of this chapter titled "Trellis Coding and Signal Structures."

V.17 training and synchronization sequences

V.17 employs two different sequences for training and synchronizing the signals between modems. A long train is used for the initial connection setup or for a connection that needs retraining. The resynchronization (resync) sequence is used for a resynchronization after a successful long train sequence. Both sequences require the modem to transmit specific sets of synchronizing signal elements, which are various but defined combinations of the signal space diagram (the constellation), which was explained in chapter 2. The sequences are sent repetitively and alternately in accordance with the V.17 requirements.

A summary of the principal operating characteristics of V.17 is provided in Table 5.4.

**TABLE 5.5 The Frequencies
Used to Represent Binary Data**

Channel 1	Channel 2
F_A=1180 Hz (0)	F_A=1850 Hz (0)
F_Z=980 Hz (1)	F_Z=1650 Hz (1)

V.21

The V.21 Recommendation is used on slow-speed systems. Some vendors offer speeds of 50 to 300 bit/s. V.21 is offered by some vendors as a small inexpensive modem that fits under a standard telephone. This recommendation uses frequency shift modulation. The modulation rate equals the bit rate. The mean frequency for channel 1 is 1080 Hz; for channel 2, it is 1750 Hz. Both synchronous or asynchronous procedures are allowed. The nominal characteristic frequencies are shown in Table 5.5. Each channel uses one of two frequencies to represent binary data.

You learned in chapter 2 about split stream modems and originate-and-answer modems. V.21 describes the operations for both of these techniques. V.21 recommends that the originating modem use the low-channel frequencies for transmitting and the high-channel frequencies for receiving. Conversely, the answer modem transmits using the high frequencies and receives on the low frequencies. Figure 5.8 shows the operations for the split stream originate-and-answer modem.

A number of the V.21 modems were built with a physical switch that enabled the terminal user to choose the call originate or answer mode. The answering modem will adjust itself to the transmit and receive frequencies

Channel 1	Channel 2
1070 Hz (0)	2025 Hz (0)
1270 Hz (1)	2225 Hz (1)

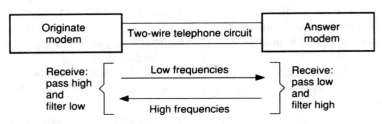

Figure 5.8 Originate-and-answer modems.

accordingly. This modem option is relatively old and seldom used today. It is quite rare to see the physical switches on such a device.

The V.21 Recommendation sometimes is compared to the Bell 103/113 modems. However, they are incompatible. The Bell 103/113 modem uses the frequencies shown in Table 5.6.

The Bell 103/113 modem often is configured to originate only at the terminal end and answer only at the computer end. This approach works well if the application on a terminal always calls the host computer. The economy results from simpler modems because they do not have to be configured with circuitry to alternate between transmitters and receivers.

A summary of the principal operating characteristics of V.21 is provided in Table 5.7.

V.22

The V.22 Recommendation, which serves as the foundation for many 1200-bit/s modems, is similar to the Bell modem 212A and is used by some of the Hayes modems.

The V.22 channels are separated by frequency division, and each channel is phase shift modulated. A 600-baud signal carries two bits per baud (dibits) using the encoded scheme shown in Table 5.8.

TABLE 5.6 The Bell 103/113 Modem Frequencies

Channel 1	Channel 2
1070 Hz (0)	2025 Hz (0)
1270 Hz (1)	2225 Hz (1)

TABLE 5.7 Principal Characteristics of V.21

Nr.	Speed	Rate	Rat	Sep	Car	H/F	S/A	Mod	B?	Sw?	Ded?	Equa	Notes
V.21	300	300	1:1	FD	1080 & 1750	F	E	FS	D	Y	O	D	YYYDY

TABLE 5.8 V.22 Encoding

Dibit values (1200 bit/s)	Bit values (600 bit/s)	Phase change (Modes i, ii, iii, iv)	Phase change (Mode v)
00	0	+ 90°	+270°
01	—	0°	+180°
11	1	+270°	+ 90°
10	—	+180°	0°

TABLE 5.9 Principal Characteristics of V.22

Nr.	Speed	Rate	Rat	Sep	Car	H/F	S/A	Mod	B?	Sw?	Ded?	Equa	Notes
V.22	1200	600	2:1	FD	1200 & 2400	F	E	PS	D	Y	PP2W	F	YYYYY
V.22	600	600	1:1	FD	1200 & 2400	F	E	PS	D	Y	PP2W	F	YYYYY

The V.22 implementations typically use V.25 dial-and-answer and V.54 loop testing procedures.

Carrier frequencies are 1200 Hz for the low channel and 2400 Hz for the high channel. V.22 permits several options in mixing 1200/600/300-bit/s and synchronous/asynchronous transmission. It also supports several start/stop formats.

Although there is some similarity between the V.22 Recommendation and the Bell 212A, the two modem types are incompatible. The reason for this incompatibility is that the Bell modem uses a Bell 103/113 frequency shift key modulation scheme that is incompatible with the V.22 modem, which, as Table 5.8 indicates, uses phase modulation. A summary of the principal operating characteristics of V.22 is provided in Table 5.9.

V.22 bis

The V.22 bis Recommendation has been implemented in many modems. The personal computer (PC) industry uses V.22 bis for medium speed, dial-up systems. The newer Hayes modems offer V.22 bis as one option.

Most V.22 bis products include unattended switching to a standby line, automatic dial and answer, adaptive equalization, and extensive diagnos-

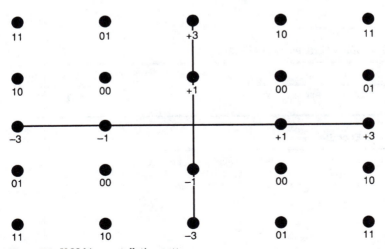

Figure 5.9 V.22 bis constellation pattern.

tics. Be aware that some vendors use different originate-and-answer frequencies.

These modems separate the channels by frequency division. Then each channel is quadrature amplitude modulated (QAM). A 600-baud signal carries four bits per baud (quadbits) with the coding scheme in Table 5.10. Figure 5.9 shows the constellation pattern for V.22 bis.

Carrier frequencies are 1200 Hz for the low channel and 2400 Hz for the high channel. Both synchronous and start/stop transmissions are supported.

The calling modem receives signals on the high channel and transmits signals on the low channel. The answering modem transmits signals on the high channel and receives signals on the low channel. Both modems must adhere to several timing conventions in the synchronous process before data can be exchanged. A summary of the principal operating characteristics of V.22 bis is provided in Table 5.11.

TABLE 5.10 V.22 bis Encoding

First two bits in quadbit (2400 bit/s) or dibit values (1200 bit/s)	From	To	Phase quadrant change
00	1	2	
	2	3	90°
	3	4	
	4	1	
01	1	1	
	2	2	0°
	3	3	
	4	4	
11	1	4	
	2	1	270°
	3	2	
	4	3	
10	1	3	
	2	4	180°
	3	1	
	4	2	

TABLE 5.11 Principal Characteristics of V.22 bis

Nr.	Speed	Rate	Rat	Sep	Car	H/F	S/A	Mod	B?	Sw?	Ded?	Equa	Notes
V.22 bis	2400	600	4:1	FD	1200 & 2400	F	E	QAM	D	Y	PP2W	F/A	DYYYY
V.22 bis	1200	600	2:1	FD	1200 & 2400	F	E	QAM	D	Y	PP2W	F/A	DYYYY

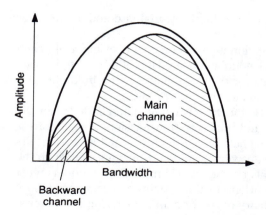

Figure 5.10 Backward channel modems.

V.23

V.23 often is implemented as a dual-rate modem with a 600- to 1200-bit/s half-duplex setup. Several vendors provide the backward channel option of 75-bit/s transmit and 600- to 1200-bit/s receive.

The backward channel option (see Figure 5.10) can be used to send a control signal, perhaps for controlling the main channel. However it is employed, the V.23 backward channel does give the capability for full-duplex transmission, albeit at very small data rate in the backward channel.

The V.23 modem frequency modulates channels in the manner shown in Table 5.12.

The majority of V.23 modems also use the V.25 dial-and-answer specification and selected test loops of V.54. V.23 permits synchronous or asynchronous modes of operation, although many vendors offer only the asynchronous version.

V.23 provides for a fallback speed of 600 bit/s. This might be needed if the transmission line is noisy and/or error prone. If the attached DTE detects unreasonable errors at the default 1200 bit/s, it informs the modem to reduce the channel speed. The signal to the modem is provided through V.24 interchange circuit 111. If this operation occurs, the V.23 modem changes its frequency from 2100/1300 Hz to 1700/1300 Hz.

TABLE 5.12 V.23 Modem Frequency Modulation

	FO	FZ (symbol 1, MARK)	FA (symbol 0, SPACE)
Mode 1: up to 600 baud	1500 Hz	1300 Hz	1700 Hz
Mode 2: up to 1200 baud	1700 Hz	1300 Hz	2100 Hz

TABLE 5.13 Principal Characteristics of V.23

Nr.	Speed	Rate	Rat	Sep	Car	H/F	S/A	Mod	B?	Sw?	Ded?	Equa	Notes
V.23	600	600	NA	NA	1300 & 1700	H	E	FM	Y	Y	O	D	DYYDY
V.23	1200	1200	NA	NA	1300 & 2100	H	E	FM	Y	Y	O	D	DYYDY

Even though the majority of V.23 modems operate at 1200 bit/s on half-duplex, it is permissible to attach this type modem to a four-wire connection that will operate in full duplex. A summary of the principal operating characteristics of V.23 is provided in Table 5.13.

V.26

The V.26 modem uses four-phase modulation with a synchronous mode of operation and also can include a backward channel of up to 75 baud.

The recommendation operates with a carrier frequency at 1800 Hz and with a modulation rate of 1200 baud. Data are encoded into pairs of bits (dibits); each dibit is encoded as a phase change relative to the immediately preceding signal. The coding can be performed in two ways; these alternatives are shown in Table 5.14, and Figure 5.11 shows the waveforms for alternative A.

The V.26 Recommendation was written specifically for private four-wire circuits. It was intended that the two separate alternatives would provide for two-way communications across the two wires and allow the modem to support 2400 bit/s in each direction. A summary of the principal operating characteristics of V.26 is provided in Table 5.15.

V.26 bis

V.26 bis uses carrier frequency, modulation, and coding according to V.26, alternative B. It also includes provisions for a reduced rate at 1200 bit/s, as well as a backward channel of up to 75 baud.

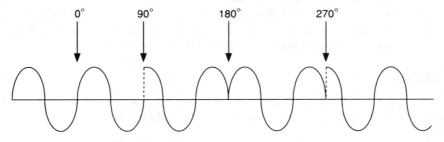

Figure 5.11 V.26, alternative A.

TABLE 5.14 V.26 Encoding

Dibit	Phase change* Alternative A	Alternative B
00	0°	+ 45°
01	+ 90°	+135°
11	+180°	+225°
10	+270°	+315°

*The phase change is the phase shift in the transition region from the center of one signalling element to the center of the following signalling element.

TABLE 5.15 Principal Characteristics of V.26

Nr.	Speed	Rate	Rat	Sep	Car	H/F	S/A	Mod	B?	Sw?	Ded?	Equa	Notes
V.26	2400	1200	2:1	4W	1800	F	S	PS	Y	N	PPMP2W	D	YYYDD

TABLE 5.16 Principal Characteristics of V.26 bis

Nr.	Speed	Rate	Rat	Sep	Car	H/F	S/A	Mod	B?	Sw?	Ded?	Equa	Notes
V.26 bis	2400	1200	2:1	NA	1800	H	S	PS	Y	Y	N	F	YYYDY
V.26 bis	1200	1200	1:1	NA	1800	H	S	PS	Y	Y	N	F	YYYDY

The recommendation operates with a carrier frequency of 1800 Hz. Refer to V.26 for information on the coding and modulation pattern. For operation at 1200 bit/s, the modem uses a two-phase shift with 0 for 90° and 1 for 270°. In this option, the modulation rate still is 1200 baud.

V.26 bis is somewhat similar to the Bell 201 modem, which also uses differential phase shift keying. The Bell 201 modem implements the alternative B option with the carrier frequency of 1800 Hz. The Bell specifications also state that the Bell 201C, which is used on a switched line, is compatible with the V.26 modem. The Bell 201B modem is designed for use on dedicated four-wire circuits, and it is compatible with V.26 using alternative B. A summary of the principal operating characteristics of V.26 bis is provided in Table 5.16.

V.26 ter

V.26 ter is a relative newcomer to the industry and has not yet seen extensive use. The principal characteristics of this modem are:

- Duplex (FDX) mode of operation is possible on switched and point-to-point leased circuits.

- Half-duplex (HDX) mode of operation is optional on switched and point-to-point leased circuits.

- Channel separation is done by echo cancellation.

- Differential phase shift modulation is performed for each channel with synchronous line transmission at 1200 baud.

- It includes a scrambler.

- It includes an equalizer.

- It includes test facilities.

- Operation with DTEs is available in the following modes: 2400 bit/s synchronous, 1200 bit/s synchronous (fallback rate), 2400 bit/s start/stop (optional), and 1200 bit/s start/stop (optional fallback rate).

- It includes an operating sequence that is intended to allow interworking with two-wire duplex 4800-bit/s modem.

The carrier frequency operates at 1800 Hz, power levels conform to V.2, and the signaling rate is 1200 baud with the 2400- and 1200-bit/s rate coded as shown in Tables 5.17 and 5.18. A summary of the principal operating characteristics of V.26 ter is provided in Table 5.19.

**TABLE 5.17 V.26 ter
Encoding at 2400 bit/s**

Dibit values	Phase change*
00	+0°
01	+90°
11	+180°
10	+270°

*The phase change is the phase shift in the transition region from the center of one signalling element to the center of the next signalling element.

**TABLE 5.18 V.26 ter
Encoding at 1200 bit/s**

Bit values	Phase change*
0	0°
1	180°

*The phase change is the phase shift in the transition region from the center of one signalling element to the center of the next signalling element.

TABLE 5.19 Principal Characteristics of V.26 ter

Nr.	Speed	Rate	Rat	Sep	Car	H/F	S/A	Mod	B?	Sw?	Ded?	Equa	Notes
V.26 ter	2400	1200	2:1	EC	1800	E	E	PS	D	Y	PP2W	E	YYYYY
V.26 ter	1200	1200	1:1	EC	1800	E	E	PS	D	Y	PP2W	E	YYYYY

V.27

Most of the V.27 implementations now use the V.27 bis or V.27 ter Recommendations. The principal characteristics of this modem are:

- It is capable of operating in an FDX or HDX mode.
- It uses differential eight-phase modulation with synchronous mode of operation.
- It can provide a backward (supervisory) channel at modulation rates up to 75 baud in each direction of transmission, the use of these channels being optional.
- It includes a manually adjustable equalizer.

Modulation occurs with an 1800-Hz carrier, the power levels conform to V.2, and the data stream is divided into three consecutive bits (tribits). Each tribit is encoded as a phase change relative to the phase of the immediately preceding signal elements. The modulation rate is 1600 baud. The tribit values are coded into the phase changes described in Table 5.20. A summary of the principal operating characteristics of V.27 is provided in Table 5.21.

TABLE 5.20 V.27 Encoding

Tribit values			Phase change*
0	0	1	0°
0	0	0	45°
0	1	0	90°
0	1	1	135°
1	1	1	180°
1	1	0	225°
1	0	0	270°
1	0	1	315°

*The phase change is the phase shift in the transition region from the center of one signalling element to the center of the next signalling element.

TABLE 5.21 Principal Characteristics of V.27 ter

Nr.	Speed	Rate	Rat	Sep	Car	H/F	S/A	Mod	B?	Sw?	Ded?	Equa	Notes
V.27	4800	1600	3:1	4W	1800	E	S	PS	Y	N	4W	M	YYYYD

V.27 bis

The characteristics of V.27 bis are very similar to V.27, except that this recommendation uses different equalization techniques and provides for 2400- or 4800-bit/s speeds. Its principal characteristics are:

- It operates in an FDX or HDX mode over four-wire leased circuits or in an HDX mode over two-wire leased circuits.

- At 4800-bit/s operation, modulation is eight-phase differentially encoded as described in Recommendation V.27.

- At a reduced rate capability of 2400 bit/s, it uses four-phase differentially encoded modulation as described in Recommendation V.26, alternative A.

- It can include a backward (supervisory) channel at modulation rates up to 75 baud in each direction of transmission.

- It includes an automatic adaptive equalizer.

Most vendors include features such as adaptive equalization, automatic retraining, scrambling, and extensive tests and diagnostics.

Modulation occurs with an 1800-Hz carrier; the power levels conform to V.2 modulation at 4800 bit/s with 1600 baud and 2400 bit/s with 1200 baud. The encoding schemes for these two techniques are described in Tables 5.22 and 5.23.

TABLE 5.22 V.27 bis Encoding at 4800 bit/s

Tribit values			Phase change*
0	0	1	0°
0	0	0	45°
0	1	0	90°
0	1	1	135°
1	1	1	180°
1	1	0	225°
1	0	0	270°
1	0	1	315°

*The phase change is the phase shift in the transition region from the center of one signalling element to the center of the next signalling element.

**TABLE 5.23 V.27 bis
Encoding at 2400 bit/s**

Dibit values	Phase change*
00	0°
01	90°
11	180°
10	270°

*The phase change is the phase shift in the transition region from the center of one signalling element to the center of the next signalling element.

TABLE 5.24 Principal Characteristics of V.27 bis

Nr.	Speed	Rate	Rat	Sep	Car	H/F	S/A	Mod	B?	Sw?	Ded?	Equa	Notes
V.27 bis	4800	1600	3:1	4W	1800	E	S	PS	Y	N	2W4W	A	YYYYD
V.27 bis	2400	1200	2:1	4W	1800	E	S	PS	Y	N	2W4W	A	YYYYD

At a fallback rate of 7200 bit/s, the same phase shifts are used, but the "amplitude bit" is not used. At a fallback rate of 4800 bit/s, the phase changes are identical to V.26 (alternative A: 0°, 90°, 180°, 270° shifts). A summary of the principal operating characteristics of V.27 bis is provided in Table 5.24.

V.27 ter

The majority of European Postal and Telegraph administrations (PTTs) support either V.27 bis or V.27 ter. These interfaces are not used much in North America. The principal characteristics of this modem are:

- It uses a data signalling rate of 4800 bit/s with eight-phase differentially encoded modulation as described in the V.27 Recommendation.

- It operates at a reduced rate capability of 2400 bit/s with four-phase differentially encoded modulation as described in Recommendation V.26, alternative A.

- It can include a backward channel at modulation rates up to 75 baud.

- It includes an automatic adaptive equalizer.

Modulation occurs at 1800 Hz, and power levels conform to V.2. The modulation patterns and codes are identical to the V.27 Recommendation. The reduced rate uses V.26, alternative A. Refer to V.27 and V.26 for these descriptions. A summary of the principal operating characteristics of V.27 ter is provided in Table 5.25.

TABLE 5.25 Principal Characteristics of V.27 ter

Nr.	Speed	Rate	Rat	Sep	Car	H/F	S/A	Mod	B?	Sw?	Ded?	Equa	Notes
V.27 ter	4800	1600	3:1	Ne	1800	H	S	PS	Y	Y	N	A	YYYYY
V.27 ter	2400	1200	2:1	Ne	1800	H	S	PS	Y	Y	N	A	YYYYY

V.29

V.29 is a widely used recommendation that is found in many North American and European products. The Bell V.29 modem is based on ITU-T V.29.

This high-speed modem operates in FDX or HDX mode using amplitude and phase modulation with synchronous transmission at 9600 bit/s. It also provides for fallback data rates of 7200 and 4800 bit/s. The carrier frequency is 1700 Hz.

This recommendation uses a modulation rate of 2400 baud and provides for three types of bit encoding for the three available speeds. At 9600 bit/s, the bits are divided into groups of four (quadbits). The first bit is used to represent amplitude, and the other three bits provide for eight possible phase shifts as shown in Table 5.26. Table 5.27 shows the operations of the first bit (Q1). Figure 5.12 shows the V.29 constellation pattern. A summary of the principal operating characteristics of V.29 is provided in Table 5.28.

Forward Error Correcting (FEC) Modems

In the past, FEC of distorted data was considered too expensive except for rather esoteric applications, such as deep space probes. With the advent of

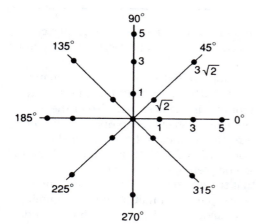

Figure 5.12 V.29 9600-bit/s signal space diagram.

TABLE 5.26 V.29 Encoding

Q2	Q3	Q4	Phase change*
0	0	1	0°
0	0	0	45°
0	1	0	90°
0	1	1	135°
1	1	1	180°
1	1	0	225°
1	0	0	270°
1	0	1	315°

*The phase change is the shift in the transition region from the center of one signalling element to the center of the next signalling element.

TABLE 5.27 The V.29 Q1 Bit

Absolute phase	Q1	Relative signal element amplitude
0°, 90°, 180°, 270°	0	3
	1	5
45°, 135°, 225°, 315°	0	$\sqrt{2}$
	1	$3\sqrt{2}$

TABLE 5.28 Principal Characteristics of V.29

Nr.	Speed	Rate	Rat	Sep	Car	H/F	S/A	Mod	B?	Sw?	Ded?	Equa	Notes
V.29	9600	2400	4:1	4W	1700	E	S	QAM	D	N	PP4W	A	YYYYD
V.29	7200	2400	3:1	4W	1700	E	S	PS	D	N	PP4W	A	YYYYD
V.29	4800	2400	2:1	4W	1700	E	S	PS	D	N	PP4W	A	YYYYD

very large-scale integration circuits (VLSI), FEC has become rather common and is found in the newer ITU-T Recommendations for voice-band modems: V.32 and V.33.

Retransmission is a time-honored compromise to the error-laden channel and has served the industry well. Nonetheless, other techniques have emerged that not only detect an error but, in many cases, correct the error without requesting a retransmission. One of these techniques, trellis coding, is highlighted here to give the reader an idea of the value of the FEC techniques. The previous section on V.17, V.32, V.32 bis, and V.33 mentioned FEC briefly. This section will complete the introduction.

Communications theory states that the greater the Hamming distance of a code, the better it can correct a corrupted data unit. A longer Hamming distance is possible by making the code more complex or changing the code ratio (ratio of FEC bits to user bits). However, the greater the number of extra (redundant) bits, the lower the user data throughput.

This creates a dilemma about how to use the existing band-limited link and still produce a greater number of bits per symbol. Moreover, the redundant bits require a denser constellation pattern, which reduces the transmission signal's immunity to noise.

However, we can reasonably pose the following scenario. Granted that the many elements involved in a data transmission often will create errors, the transmission or signal always starts at a known value and the value is confined within certain limits. Also, suppose a method is devised whereby the signal (derived and coded from the user data bit stream) is allowed to assume only certain characteristics (states) on the line. Furthermore, suppose the user bits are interpreted such that only certain of the states are allowed to exist from prior states. This means that the transmitting device accepts a series of user bits and develops additional, yet restricted, bit patterns from these bits. Moreover, the previous user bit pattern (called a *state*) is allowed to assume only certain other bit patterns (states). Other states are invalid and are never transmitted.

The transmitter and receiver are programmed to understand the allowable states and the permissible state transitions. If the receiver receives states and state transitions (because of channel impairments) that differ from predefined conventions, it is assumed an error has occurred on the circuit.

However, trellis coding goes further. Because, by convention, the transmitter and the receiver know the transmission states and the permissible state transitions, receiver analyzes the received signal and makes a "best guess" as to what state the signal should assume. It analyzes current states, compares them to previous states, and makes decisions as to the most relevant state. In effect, it uses a path history to reconstruct damaged bits.

Several modem manufacturers claim that their trellis coded modems can improve the error performance of a communications channel by two to three orders of magnitude.

V.32

This recommendation (which initially was published in the Red Book) has created considerable interest because V.32 modems can operate at duplex rates of up to 9600 bit/s on a dial-up, two-wire line. Many manufacturers now offer a V.32 modem. Several modems are available that use TCM on 19.2-kbit/s links, although as of this writing the ITU-T does not publish anything beyond the 14.4-kbit/s data transfer rate. The principal characteristics of V.32 are:

- It provides a duplex mode of operation on switched and two-wire, point-to-point leased circuits.

■ Any combination of the following data signalling rates can be implemented in the modems: 9600 bit/s synchronous, 4800 bit/s synchronous, and 2400 bit/s synchronous (for further study).

Refer back to "Trellis Coding and Signal Structures" earlier in this chapter for further explanation of V.32.

A summary of the principal operating characteristics of V.32 is provided in Table 5.29. The constellation pattern for a V.32 was illustrated earlier.

V.32 bis

V.32 bis is quite similar to V.32. Its principal difference is its support for 14,400 bit/s rates. It is compatible with V.32 modems for 9600 and 4800 bit/s transfer rates. Once again, you should refer back to "Trellis Coding and Signal Structures" for further information on V.32 bis. A summary of the principal operating characteristics of V.32 bis is provided in Table 5.30.

V.33

V.33 is designed to operate over point-to-point, four-wire leased telephone-type circuits. It operates in a manner similar to the V.32 modem except that it encodes a redundant bit and six information bits for mapping into signal elements for a 2400-baud signal. With the exception of the description in Figure 5.5, it operates with the same logic as the V.32 modem. A summary of the principal operating characteristics of V.33 is provided in Table 5.31.

TABLE 5.29 Principal Characteristics of V.32

Nr.	Speed	Rate	Rat	Sep	Car	H/F	S/A	Mod	B?	Sw?	Ded?	Equa	Notes
V.32	9600	2400	4:1	EC	1800	F	S	QAM	D	Y	PP2W	A	YYYYY
V.32	9600	2400	5:1	EC	1800	F	S	TCM	D	Y	PP2W	A	YYYYY
V.32	4800	2400	2:1	EC	1800	F	S	QAM	D	Y	PP2W	A	YYYYY

TABLE 5.30 Principal Characteristics of V.32 bis

Nr.	Speed	Rate	Rat	Sep	Car	H/F	S/A	Mod	B?	Sw?	Ded?	Equa	Notes
V.32 bis	14400	2400	7	EC	1800	F	S	TCM	D	Y	PP2W	A	YDDYY
V.32 bis	12000	2400	6	EC	1800	F	S	TCM	D	Y	PP2W	A	YDDYY
V.32 bis	9600	2400	5	EC	1800	F	S	TCM	D	Y	PP2W	A	YDDYY
V.32 bis	7200	2400	4	EC	1800	F	S	TCM	D	Y	PP2W	A	YDDYY
V.32 bis	4800	2400	2	EC	1800	F	S	QAM	D	Y	PP2W	A	YDDYY

TABLE 5.31 Principal Characteristics of V.33

Nr.	Speed	Rate	Rat	Sep	Car	H/F	S/A	Mod	B?	Sw?	Ded?	Equa	Notes
V.33	14400	2400	7:1	EC	1800	F	S	TCM	D	Fu	PP4W	A	YYYYN

What does FEC hold in the future for data communications systems and networks? The use of the technique will lead to changes to much of our software and hardware that concerns itself with the transmission and reception of data in our currently existing systems. For example, according to some vendors, TCM reduces the error rate by three orders of magnitude. For this reason, FEC is becoming pervasive, and data link controls (line protocols) will likely undergo significant changes.

Summary of the ITU-T Voice-Band Modems

Table 5.32 summarizes the V Series voice-band modems. This table is an abbreviated compilation of the tables in this chapter that listed the characteristics of each of the V Series modems.

The Next Generation of V Series Modems: V.34

The ITU-T recently has approved a higher speed modem, designated V.34. I have delayed the rewrite of this book for several months in the hopes that specifications would be completed. This part of the book will explain V.34, even though the specification has not yet been printed for the public.

The standard also is known in the industry as *V.fast*. Its architecture is based on trellis coded modulation and QAM techniques that were discussed early in this chapter. Its most noticeable difference is its speed with a signaling rate of 28.8 kbits/s. Its signaling rate can be 2400 baud or even 3400. Currently, the V.34 Recommendation specifies a data rate of 28.8 kbits/s.

The V.fast-V.32 terbo debate

Due to the slowness of the V.34 implementation, several vendors have formed an association to produce their own specification, which is known in the industry as *V.32 terbo*. It is similar to V.32 bis and uses the same coding scheme but has more constellation points (to either to 256 or 512), which allows the signaling rate of 16.8 kbits/s or 19.2 kbits/s.

The different modem manufacturers have lined up either behind V.fast or V.32 terbo. Some of the vendors believe that it is counterproductive to develop an interim standard and that the formal standard should be available in a short period of time. Other vendors have grown weary with this slow process. They state that a standard is needed immediately and that an in-

TABLE 5.32 ITU-T V Series Voice-Band Modems

Nr.	Speed	Rate	Rat	Sep	Car	H/F	S/A	Mod	B?	Sw?	Ded?	Equa	Notes
V.21	300	300	1:1	FD	1080 & 1750	F	E	FS	D	Y	O	D	YYYDY
V.22	1200	600	2:1	FD	1200 & 2400	F	E	PS	D	Y	PP2W	F	YYYY
V.22	600	600	1:1	FD	1200 & 2400	F	E	PS	D	Y	PP2W	F	YYYY
V.22 bis	2400	600	4:1	FD	1200 & 2400	F	E	QAM	D	Y	PP2W	F/A	DYYYY
V.22 bis	1200	600	2:1	FD	1200 & 2400	F	E	QAM	D	Y	PP2W	F/A	DYYYY
V.23	600	600	NA	NA	1300 & 1700	H	E	FM	Y	Y	O	D	DYYDY
V.23	1200	1200	NA	NA	1300 & 2100	H	E	FM	Y	Y	O	D	DYYDY
V.26	2400	1200	2:1	4W	1800	F	S	PS	Y	N	PPMP2W	D	YYDD
V.26 bis	2400	1200	2:1	NA	1800	H	S	PS	Y	Y	N	F	YYDY
V.26 bis	1200	1200	1:1	NA	1800	H	S	PS	Y	Y	N	F	YYYDY
V.26 ter	2400	1200	2:1	EC	1800	E	E	PS	D	Y	PP2W	E	YYYY
V.26 ter	1200	1200	1:1	EC	1800	E	E	PS	D	Y	PP2W	E	YYYY
V.27	4800	1600	3:1	4W	1800	E	S	PS	Y	N	4W	M	YYYD
V.27 bis	4800	1600	3:1	4W	1800	E	S	PS	Y	N	2W4W	A	YYYD
V.27 bis	2400	1200	2:1	4W	1800	E	S	PS	Y	N	2W4W	A	YYYD
V.27 ter	4800	1600	3:1	Ne	1800	H	S	PS	Y	Y	N	A	YYYY
V.27 ter	2400	1200	2:1	Ne	1800	H	S	PS	Y	Y	N	A	YYYY
V.29	9600	2400	4:1	4W	1700	E	S	QAM	D	N	PP4W	A	YYYD
V.29	7200	2400	3:1	4W	1700	E	S	PS	D	N	PP4W	A	YYYD
V.29	4800	2400	2:1	4W	1700	E	S	PS	D	N	PP4W	A	YYYD
V.32	9600	2400	4:1	EC	1800	F	S	QAM	D	Y	PP2W	A	YYYY
V.32	9600	2400	5:1	EC	1800	F	S	TCM	D	Y	PP2W	A	YYYY
V.32	4800	2400	2:1	EC	1800	F	S	PS	D	Y	PP2W	A	YYYY
V.33	14400	2400	7:1	EC	1800	F	S	TCM	D	Fu	PP4W	A	YYYN

terim standard indeed will do no harm but will move the industry forward quickly to a common approach.

You should carefully check the vendors' modems that claim to use V.34. Most vendors that are not V.34 compatible have announced plans to upgrade/migrate to the final V.34 specification.

Summary

The voice-band modem recommendations that were published by the ITU-T have become *the* standard from which vendors now design their modem products. In the past, we learned that the Bell modems were considered to be *de facto* standards in certain parts of the world. However, in the past few years, the ITU-T V Series modems have become the dominant standard. The voice-band modems describe data rates up to and including 14.4 kbit/s. Beyond this speed there is no recognized international standard as of this writing. You also learned that vendors have increasingly implemented the higher-speed V.32 and V.33 modems because of their forward correcting capabilities.

6

Wide-Band Modems

The ITU-T does not devote much space to wide-band modems. Indeed, as the world moves to digital communications lines and digital networks, it is expected that wide-band analog modems will see less use. This chapter is rather short, which reflects the brevity of the ITU-T V Series Recommendations dealing with these machines.

Figure 6.1 depicts the structure of the ITU-T V wide-band modems. As the figure illustrates, the ITU-T publishes recommendations for three wide-band modems:

- V.35: Data Transmission at 48 kbit/s using 60- to 108-kHz group band circuits (NLIF).

- V.36: Modems for synchronous data transmission using 60- to 108-kHz group band circuits.

- V.37: Synchronous data transmission at a data signalling rate higher than 72 kbit/s using 60- to 108-kHz group band circuits.

- V.38: A 48/56/64 kbit/s data circuit terminating equipment that is standardized for use on digital point-to-point leased circuits.

V.35

V.35 is an old standard (1976) that is designed for duplex operation on leased circuits. It is no longer in force (NLIF) and, therefore, is not shown in Figure 6.1. It is a unique standard in that some of its V.24 interchange cir-

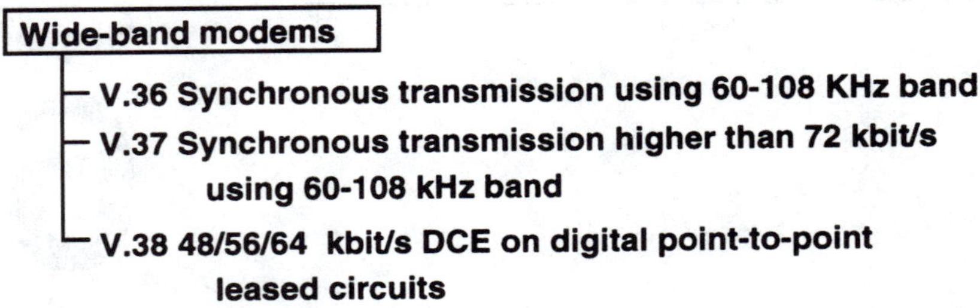

Wide-band modems

— **V.36 Synchronous transmission using 60-108 KHz band**

— **V.37 Synchronous transmission higher than 72 kbit/s using 60-108 kHz band**

— **V.38 48/56/64 kbit/s DCE on digital point-to-point leased circuits**

Figure 6.1   V Series wide-band modems.

cuits are defined with the V.28 electrical specification, and the others use something similar to (but incompatible with) V.11.

In typical installations, a mark is represented by +0.55 volt ±20%, and a space is represented by –0.55 volt ±20%. V.35 is a balanced double signal interchange and uses two wires for differential signaling on the clock and data interchange circuits. When the A lead is positive relative to the B lead, a space is represented. When the B lead is positive relative to the A lead, a mark is represented.

A specific connector is not specified in the V.35 Recommendation, although the V.24 interchange circuits are defined (as in Table 6.1). Most commercial implementations use the 34-pin Winchester connector, which is illustrated in Figure 6.2. For this connector, the V.35 pin layout is configured as shown in Table 6.2.

This modem operates at 48 kbit/s. The data signal is translated to the 60- to 108-kHz band as a side-band suppressed carrier AM signal. The carrier frequency operates at 100 kHz. The interface cable is a balanced twisted pair wire.

TABLE 6.1 V.35 Interchange Circuits

No.	Function
102*	Signal Ground or Common Return
103	Transmitted Data
104	Received Data
105*	Request to Send
106*	Ready for Sending
107*	Data Set Ready
109*	Data Channel Receive Line Signal Detector
114	Transmitter Signal Element Timing
115	Receiver Signal Element Timing

*Circuits conform to V.28. The other circuits are a balanced pair.

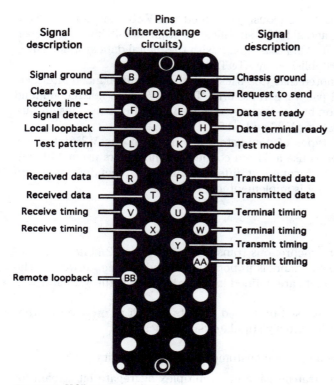

Figure 6.2 V.35 connector (male).

TABLE 6.2 V.35 Connectors and Pinout

34-pin —	A	B	V.35 circuit	Function
A			FG	Frame (protective ground)
B			SG	Signal ground
	P	S	SD	Send data
	R	T	RD	Receive data
C			RTS	Request to send
D			CTS	Clear to send
E			DSR	Data set ready
H			DTR	Data terminal ready
F			RLSD	Received line signal detector
	U	W	SCTE	Serial clock transmit ext
	Y	AA	SCT	Serial clock transmit
	V	X	SCR	Serial clock receive
J				Local loopback
BB				Remote loopback
K			LT	Test mode
L				Test pattern

It is commonly (and erroneously) assumed that V.35 is specified as a 56-kbit/s interface. The actual ITU-T specification stipulates 48 kbit/s. However, many systems operate V.35 at 56 kbit/s with the digital data system (DDS) offering originally established by AT&T.

It also should be noted that V.35 uses both the V.28 unbalanced interface, as well as balanced pair (see Table 6.1). The ITU-T does not recommend that this specification be used on any new designs. Alternative designs are to use Recommendations V.36 and V.37.

Finally, different types of connectors are used for the V.35 interface. Some implementations use a 34-pin connector, and others use a 449 connector. In essence, V.35 comes in different varieties, depending on where (geographical region) it is implemented.

V.36

V.36 is designed for synchronous transmission at 48 to 72 kbit/s. It is employed on multiplexed channels in both public and private networks. The V.24 interchange circuits are defined in accordance with the V.10 and V.11 Recommendations.

This modem type is used on leased circuits as well as on several other kinds of facilities. V.36 actually stipulates six applications:

- It will transmit data between customers on leased circuits.

- It will provide the transmission of a multiplex aggregate bit stream for public data networks.

- It provides for the extension of a pulse code modulation (PCM) channel at 64 kbit/s over analog facilities.

- It provides for transmission of a common channel signalling system for telephony and/or public data networks.

- It provides for the extension of a single-channel-per-carrier (SCPC) circuit from a satellite earth station.

- It allows the transmission of a multiplex aggregate bit stream for telegraph and data signals.

Application 1. The recommended data signalling rate for international use is 48 kbit/s. For certain national applications or with bilateral agreement, the following data signalling rates are applicable: 56, 64, and 72 kbit/s.

Applications 2, 3, and 4. For these applications, the recommended data signalling rate is synchronous at 64 kbit/s. For synchronous networks requiring end-to-end transmission of both the 8- and 64-kHz timing together with a data rate at 64 kbit/s, a data signalling rate of 72 kbit/s on the line is suggested.

Because a 64-kbit/s PCM channel can be used with these application choices, V.36 describes methods of using the basic rate E bit to provide alignment and housekeeping information on this link.

Application 5. The recommended data signalling rate for international use is synchronous at 49 kbit/s. For certain national applications or with bilateral agreement, the data signalling rate of 56 kbit/s is applicable.

Application 6. The recommended data signalling rate is synchronous at 64 kbit/s.

The V.36 60- to 108-kHz band corresponds to a single side-band signal using a carrier of 100 kHz ±2 Hz. Filters (which were introduced in chapter 2) are used to obtain these bands. In addition, V.36 cites V.1 to obtain a binary 1 with tone on and a binary 0 with tone off.

V.36 permits interchange circuits conforming to V.10 or V.11. It uses the ISO connector number 4902. Table 6.3 lists the V.24 interchange circuits that are used with this recommendation. V.36 requires that circuits 103, 104, 113, 114, and 115 use generators and receivers that adhere to V.11. Circuits 105, 106, 107, and 109 require that the generators use V.10, or alternately V.11. All receivers for these circuits will use V.10 category 1 or V.11 without any termination. For all other circuits, V.10 is used with the receivers except that they are configured for V.10 category 2.

V.37

V.37 was introduced with the Red Book. It operates with single side-band modulation at rates of up to 168 kbit/s. This modem also is used on leased

TABLE 6.3 V.36 Interchange Circuits

No.	Function
102	Signal Ground or Common Return
102a*	DTE Common Return
102b*	DCE Common Return
103	Transmitted Data
104	Received Data
105	Request to Send
106	Ready for Sending
107	Data Set Ready
109	Data Channel Receive Line Signal Detector
113	Transmitter Signal Element Timing (DTE source)
114	Transmitter Signal Element Timing (DCE source)
115	Receiver Signal Element Timing (DCE source)
140	Loopback Test
141	Local Loopback
142	Test Indicator

* Circuits 102a and 102b are required when V.10 is used.

circuits. The only group reference pilot frequency that can be used in conjunction with this modem is 104.08 kHz. Its principal characteristics are:

- It transmits any type of high-speed synchronous data in duplex, constant carrier mode, on four-wire (60 to 108 kHz) group band circuits.
- It has primary data signalling rates up to 144 kbit/s.
- It includes an automatic adaptive equalizer.
- It includes partial response pulse amplitude single side-band signalling and modulation.
- It has optional inclusion of an overhead-free multiplexer combining existing data signalling rates.
- It has an optional voice channel.

V.38

This specification was introduced in 1993 to define the characteristics of point-to-point leased circuits utilizing digital technology. It defines the operations only between the DCEs with stipulation for the use of V.24 between the DCE and DTE. The signalling rate is either 56 or 64 kbit/s. Although V.110 can be employed for rate adaptation of 48 kbit/s as well as slower speeds ranging from 2400 bit/s up to 19200 bit/s, the intent of the recommendation is to define the signaling characteristics on circuits other than ISDN channels. The electrical characteristics of the interface conform to either V.10 or V.11, and the connectors and pin assignments are specified in ISO 4902 or ISO 2110. The V.24 circuits are employed as depicted in Table 6.4. Test loops are used in accordance with V.54.

There is not too much more to be said on V.38. Obviously, you must understand the other V specifications that are described in this book to understand how V.38 operates.

Summary

Although no longer in force, V.35 is the most widely used wide-band interface of the V Series wide-band recommendations. It is implemented in different "flavors," depending on the country and part of the world in which the modem is installed. Again, be aware that V.35 is designated as an analog interface, but some manufacturers have modified the standard to provide for digital transmission schemes as well.

Table 6.4 V.38 use of V.24

Pin Number	Interchange circuit
102	Signal ground or common return
102a	DTE common return
102b	DCE common return
103	Transmitted
104	Received data
105	Request to send
106	Ready for sending
107	Data set ready
108/2	Data terminal
109	Data channel received line signal detector
113	Transmitter signal element timing (DTE source)
114	Transmitter signal element timing (DCE source)
115	Receiver signal element timing (DCE source)
140	Loopback/Maintenance test
141	Local loopback
142	Test indicator

7

Error Control

Until recently, the V Series Recommendations did not contain much material on procedures for dealing with data errors. In the past, the vast majority of error control measures were implemented in layers other than the physical layer. As you learned earlier, most of the V Series Recommendations reside in the physical layer.

With the decreased costs and increased power of very large-scale integration (VLSI) technology, it has become feasible to place error control operations in hardware at the physical layer. My purpose in this chapter is to review, in a general way, two older error control recommendations: V.40 and V.41. However, the emphasis is on two relatively new error control recommendations, which have received a great deal of attention in the industry and now are offered in a wide variety of products: V.42 and V.42 bis.

The titles of the V Series Recommendations dealing with error control are (see also Figure 7.1):

- V.40: Error indication with electromechanical equipment (NLIF)

- V.41: Code-independent error-control system

- V.42: Error-correcting procedures for DCEs using synchronous-to-asynchronous conversion

- V.42 bis: Data compression procedures for DCEs using error correcting procedures

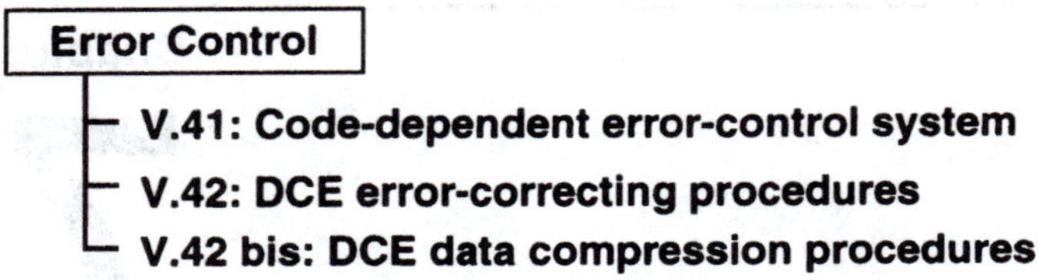

```
┌─────────────────────┐
│  Error Control      │
└─────────────────────┘
    ├─ V.41: Code-dependent error-control system
    ├─ V.42: DCE error-correcting procedures
    └─ V.42 bis: DCE data compression procedures
```

Figure 7.1 The V Series error control recommendations.

V.40

V.40 is a rather antiquated specification consisting of one-half page. It was published in 1968 and deals with the use of perforated tape and error counting techniques on older equipment. It is no longer in force (NLIF).

V.41

V.41 also is an old recommendation that was published initially in 1968, then amended in 1972. It stipulates that error control can be implemented in either the data terminal equipment (DTE) or the data circuit-terminating equipment (DCE). It is designed for error detection on the older, lower-speed modems that operated from 200 to 4800 bit/s. It was intended principally for use in the V.23 modems.

V.42

The V.42 specification has aroused considerable interest in the user community because it addresses two major problems that have existed since the use of asynchronous devices (and especially asynchronous personal computers) began: an asynchronous-to-synchronous conversion protocol and a more sophisticated error-detection process for asynchronous systems than exists with simple echo checks and parity checks.

V.42 is designed to perform asynchronous-to-synchronous conversion (using V.14) as well as error detection and retransmission of damaged data. V.42 is the culmination of several years of efforts by ITU-T, working in conjunction with users and vendors.

The approach of V.42 is shown in Figure 7.2, which illustrates that error detection and correction have been moved to the physical layer. This means that asynchronous link protocols using parity and echo checks can be replaced by V.42.

V.42 includes the following services:

- It uses a link control protocol for the sequencing and flow control of the traffic.

- It uses a cyclic redundancy check for error detection.
- It uses V.14 for asynchronous-to-synchronous conversion operations.
- Transmission of synchronous data is provided on the communications link.

A V.42 error detecting DCE contains four major components, which are illustrated in Figure 7.3:

- A signal converter
- An error control function
- A control function
- A V.24 interface

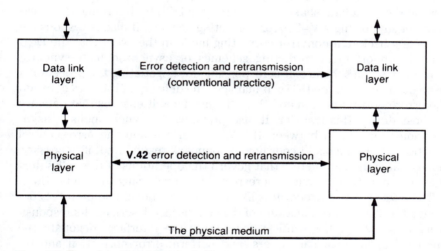

Figure 7.2 Error detection at the physical layer.

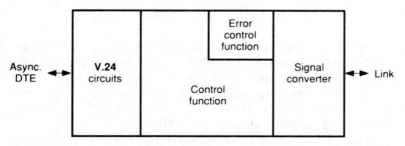

Figure 7.3 The error control modem (V.42).

The V.42 Recommendation establishes the error control function as the service provider and the control function as the service user. These terms are used in the context of the OSI Model.

The V.24 interface provides the connection to the DTE. Data are exchanged on the interface in a start/stop format. The signal converter is the interface to the telephone line. It is designed to operate on a two-wire, point-to-point leased circuit (although I see no reason that the specification could not be used on dial-up circuits, with V.25, V.25 bis, or EIA 366 dial-up procedures established as an initial handshake). The control function coordinates and controls the other DCE functions. The error control function is responsible for the error correction protocol.

Responsibilities of the control and error control functions

The *control function* is the "operating system" of V.42. It is responsible for conducting a handshake with the remote DCE to determine if it is capable of supporting a V.42 error-correcting scheme. It also is responsible for falling back to a nonerror-correcting mode in the event that the DCE cannot support error-correcting schemes or asynchronous-to-synchronous conversion schemes. It is responsible for negotiating the procedures to be used between the DCEs. It manages the delivery of the data between the error control function and the V.24 interface. It ensures that data are not lost during this transfer. It also provides the synchronous-to-asynchronous conversion between the V.24 interface and the error-control function. It provides and coordinates loopback testing and, if necessary, renegotiates the parameters that govern the operations of the DCEs during the connection. Finally, it is responsible for releasing the connection.

The *error-control function* is like most conventional line protocols developed by ITU-T and is unaware of the actions just described. Its responsibility is to deliver the traffic safely across the interface, negotiate the link-level operational parameters, and perform error correction and retransmission of corrupted data. It is responsible for responding to loopback testing and for releasing the connection as dictated by the control function.

LAPM

V.42 implements a link control protocol called *link access procedure for modems* (LAPM). It is based on the high-level data link control (HDLC) family of protocols and also was derived from the link access procedure on the D channel (LAPD) specification, which is part of the ITU-T integrated services digital network (ISDN) Recommendations. Figure 7.4 shows the format for the LAPM exchange identification (XID) frame.

LAPM uses the HDLC balanced asynchronous (BA) class. It makes use of HDLC functional extensions BA 1, 2, 4, 7, and 10. This mode is identical to

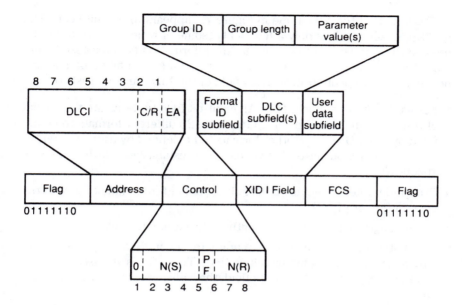

Flag: Delineates beginning and ending of frame
Address: Identifies V.24 interfaces
Control: Used for sequencing, flow control
Information: User data or control headers
FCS: Frame check sequence (for error checking)
N(S): Sending sequence number
N(R): Receiving sequence number
P/F: The poll or final bit
C/R: Command/response bit
EA: Address field extension
DLCI: Data link connection identifier

Figure 7.4   The LAPM XID frame.

LAPD. It also permits an optional procedure using HDLC functional extensions 3, 12, and 14. The reader can refer to appendix A for a discussion of
HDLC and its functional extensions.

The principal difference between LAPM and a conventional HDLC implementation is the use of the address field, which consists of the data link
identifier, the command/response (C/R) bit, and the address extension bit.
The C/R bit identifies the frame as either a command or response. The data
link control identifier (DLCI) value is used to transfer information between
the X.24 interfaces. Currently, DLCI is set to 0 to identify a DTE/DTE interface. Other values are permitted within the limits defined in the recommendation. The address extension bit can be set to 1 to designate another
octet for DLCI.

The format of the XID frame in Figure 7.4 appears to be quite complex, but it basically involves only a few fields (some of which are yet to be defined in the standard). I focus on the XID information field because it contains the parameters that are used to negotiate a number of features between the two peer entities. The I field of the XID frame consists of:

- Format identifier subfield (FI): An eight-bit field to designate the format of the I field. It has been set up to allow 128 different formats to be defined by the ISO and 128 other formats to be defined by users. Currently, only a "general-purpose" format has been established (with a value of 10000010).

- Group identifier: Specifies a user data subfield in conjunction with the general format FI.

- Group length (GL): Gives the length of the parameter field.

- Parameter field: Composed of parameter identifiers (PI), parameter lengths (PL), and parameter values (PV). These fields are used to establish the maximum length of the I field, the transmit/receive window sizes, etc.

- User data subfield: At the present time, the only values that are defined in this field are 1 = manufacturer's ID not assigned by ITU-T and 0 = manufacturer's ID assigned by ITU-T.

V.42 includes an appendix that defines an alternate procedure to LAPM. This procedure is based on the Microcom MNP protocol. This appendix was included by the ITU-T because of the prevalence of MNP in many personal computer systems. The ITU-T engineers state that the alternate protocol is fully compatible with MNP. ITU-T also states that it does not intend to provide updates and enhancements to this appendix.

V.42 primitives

The communications between the control function and the error control function occur through OSI-type primitives. These primitives are used by the control function to direct the actions of the error control function and for the error control function to inform the control function of its activities. The V.42 primitives are summarized in Table 7.1.

Establishing communications between the DCEs

The establishment of the V.42 session has two phases: the detection phase and the protocol establishment phase.

Detection phase. The detection phase is performed transparently by the control function; the user DTE and the error-control function are not aware

TABLE 7.1 V.42 Primitives

Primitive	Function
L-ESTABLISH	Establishes a connection between error-correcting entities
L-DATA	Transfers data
L-RELEASE	Releases the connection between the entities
L-SIGNAL	Sends a break signal
L-SETPARM	Establishes or negotiates parameters for the session
L-TEST	Sets up a look-back test between entities

of the operation. It uses conventional dial-and-answer techniques found in many modems today. It does not define the specific dial-and-answer standard but defers to the particular modulation recommendations.

Naturally enough, the originating DCE becomes the originator, and the answering DCE becomes the answerer. The roles assumed during this carrier handshake depend on the specific modem standards.

Several detection operations are shown in Figure 7.5. The originator begins the handshaking operations when the dial-and-answer procedures are finished. This means that the circuits ready-for-sending (RFS) and received-signal detector (RSD) are in the on condition. The originator control function then sends out a special bit pattern that is called the *originator detection pattern* (ODP). The pattern is an ASCII DC1 with even parity followed by 8 to 16 binary 1s, then another DC1 with odd parity followed by 8 to 16 1s. This is used by the receiving DCE as a "hello." If the originating DCE does not receive a response to this bit pattern, it might or might not fall back to a nonerror-correcting mode. Whatever the case might be, it assumes that V.42 capability does not exist in the remote DCE.

The remote DCE, upon receiving the ODP, becomes the answerer. As shown in Figure 7.5, if it is "smart" enough to understand that the originating DCE is capable of error-correcting operations, it sends back an answer-detection pattern (ADP) to indicate if V.42 is supported and if an error-correcting protocol is not desired. Let's assume that it transmits back the pattern indicating that V.42 is supported. The two DCEs now have completed the detection phase, and the second phase is entered. Figure 7.5 also shows the operations if the answerer does not respond initially but does respond to an MNP request and also if the answerer never responds, in which case the detection phase is not completed.

Protocol establishment phase. After the completion of the detection phase, the protocol establishment phase is entered to establish the logical link and negotiate the parameters for the operation. At this point, the V.42 primitives are invoked.

The protocol establishment phase is shown in Figure 7.6. This figure shows that 18 discrete steps are involved. You should keep in mind that this is only one scenario, showing a typical connection establishment and transfer of data along with some negotiation of parameters. Several other scenarios are available. The following discussion summarizes the operations depicted in Figure 7.6.

The first six steps are used to establish the logical connection between the two DCEs, as well as the link-layer connection. The transmitting control function gets things started by sending the L-ESTABLISH request primitive to its error-control function. This primitive is mapped to an LAPM set asynchronous balance mode extended command frame. This frame is transferred to the remote ECF, which maps the frame to a link establishment indication primitive (L-ESTAB ind). In step 4, the remote control function responds with a link establishment response primitive (L-ESTAB res). Its

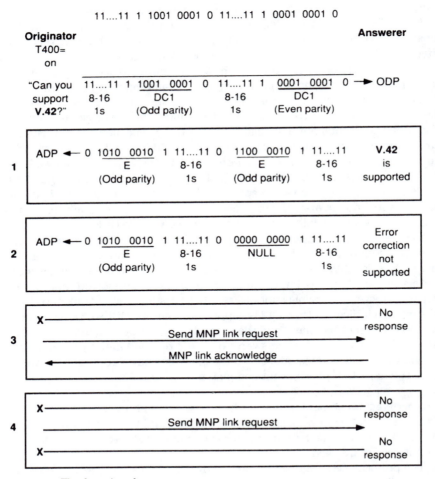

Figure 7.5 The detection phase.

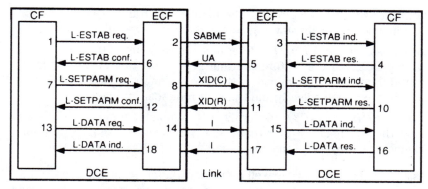

Figure 7.6 The V.42 protocol establishment phase.

supporting ECF maps this primitive to an LAPM unnumbered acknowledgment frame (UA). This frame is transmitted across the link, where it is received by the local ECF. The UA frame then is mapped to the link establishment confirm primitive (L-ESTAB conf). These operations complete the initial handshaking between the two DCEs.

Next, the control function might choose to negotiate parameters for the session. Figure 7.6 shows that it sends an L-SET parameter request primitive. Its ECF uses this primitive to create an XID LAPM frame. The parameters in this frame are used by the receiving DCE to determine several operating constraints to be used during the session. As you can see from the figure, the receiving ECF receives the XID frame and maps it to an L-SET parameter indication primitive.

In step 10, the remote control function responds with L-SET parameter response, which is mapped to the XID response frame in step 11. To complete this part of the process, the frame is sent to the local ECF, which maps the XID frame to the L-SET parameter confirm primitive.

The parameters that are negotiated in steps 9 and 10 are as follows:

- *Detection phase timer (T400).* Determines the amount of time that the ECF will wait for the originator detection pattern (ODF) or the answer detection pattern (ADF). The default value is 750 ms.

- *Detection timer (T401).* Determines the amount of time that the DCE will wait for an acknowledgment from the remote station before it resorts to remedial actions.

- *Maximum retransmissions (N400).* Determines the number of times that the DCE will retransmit a frame requiring a response.

- *Maximum number of octets in an I field (N401).* Determines the number of octets that can be carried in the I field of an I frame, an XID frame, a UI frame, or a test frame. The default is 128 octets.

- *Window size (k).* Determines the number of I frames that can be outstanding before an acknowledgment is required. The default value is 15.

- *Reply delay timer (T402).* Determines the amount of time that the receiving DCE will wait before transmitting a reply to a previously transmitted frame. This timer ensures that the transmitting DCE's T401 timer does not expire needlessly and transmit redundant frames.

- *N activity timer (T403).* Determines the amount of time that a DCE will allow the channel to be idle; that is to say, it is not transmitting any valid frames. The purpose of this timer is to detect faults as early as possible.

Assuming that all has gone well in the negotiation of the session parameters, the remaining operations in the figure depict the transfer of data using the data primitives and the LAPM information frames.

You should note that the operating parameters deal with activities that should remain transparent to the end user. Systems personnel are quite interested in these parameters, but again they are beyond, and should be kept beyond, the end user. Indeed, this standard does not stipulate (with the exception of the manufacturer's ID) the values that must reside in the I field.

For data transfer, V.42 simply carries the user information across the transmit lead of V.24 and encapsulates it into the I field of LAPM for safe transport across the link to the receiving DCE. You learned earlier in this chapter that this is the principal function of V.42; that is to say, it is an error correcting protocol.

V.42 bis

During the work on V.42, which was completed in 1988, the ITU-T study group XVII decided that a data compression enhancement was needed to further the performance of error correcting modems. Consequently, a number of existing schemes were analyzed, notably British Telecom's BTLZ, Hayes' System, Microcom's MNP5 and MNP7, and the ACT Formula. The decision was made to use the BTLZ algorithm, which I'll examine shortly. V.42 bis was not published in the Blue Book but has undergone rapid development and, as of this writing, is now an approved standard.

Compression attempts to make better use of a potentially overused resource: the communications channel. The following discussion illustrates this point.

Practically all of the symbols that are generated and used by computers are comprised of a fixed number of bits coded to represent a character. The codes (for example, ASCII) have been designed in fixed-length format because computers require a fixed number of bits in a code to efficiently process data. Most machines use octet (eight-bit) alignment.

The fixed-length format requirement means that all transmitted characters are of equal length, even though the characters are not transmitted with equal frequency. For example, characters—such as vowels, blanks, and numbers—are used and transmitted more often than consonants and characters such as a question mark. This practice can lead to considerable link inefficiency.

One widely used solution to code and channel utilization inefficiency is to adapt a variable-length value (code) to represent the fixed-length characters. In this manner, the most frequently transmitted characters are compressed; they are represented by a unique bit set smaller than the conventional bit code. This data compression technique can result in substantial savings in communications costs.

To gain an understanding for the need for data compression capabilities, consider that a normal page of text contains 1920 characters. Assuming that an eight-bit code is used to represent a character, a total of 15,360 bits will be transferred across a communications link (this number is significantly understated because it does not include start/stop bits and other control functions). Therefore, the use of a conventional V.22 bis 2400-bit/s modem can cause significant throughput and response time problems for certain types of transmissions.

The V.42 bis data compression recommendation has a compression ratio of 3:1 to 4:1 (based on the use of ASCII text). The recommendation requires approximately 3K of memory. The dictionary size for the characters and strings can be as little as 512 bytes and up to 2048 bytes. The designer must analyze the tradeoffs between smaller and larger dictionaries and must consider the fact that the larger dictionaries provide better performance with higher compression ratios even though they consume more memory. You should keep in mind that V.42 bis permits the same software to be used on different size dictionaries.

As depicted in Figure 7.7, the V.42 bis model is quite similar to the V.42 model. Logically enough, the only additional functional module is the data compression function.

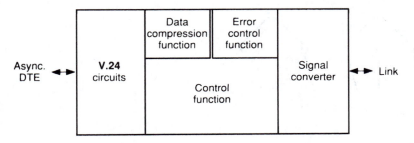

Figure 7.7 The V.42 bis model.

The V.42 bis compression algorithm works on strings of characters. It does not work on character substitution but encodes a string of characters with a fixed-length code word. The system also uses a dictionary to store frequently occurring character strings along with a code to represent these character strings.

The dictionary is a hierarchical tree structure, as illustrated in Figure 7.8. It is organized around a root and branch structure. A *root* has no higher entry and also is called a *parent* to the lower levels of the tree. Each root begins with one character in a code set (for example, the eight-bit ASCII code). Therefore, a dictionary could contain as many as 256 trees if an eight-bit character is used ($2^8 = 256$).

The tree consists of *branches* that connect nodes. Each node is represented by a character. Therefore, a tree contains a known set of character strings that begin with the character situated at the root. Each node is associated with a code word. The V.42 bis Recommendation requires that the encoder and decoder use equivalent code words for the nodes to permit easy decoding of a string.

Figure 7.8 shows an example of the dictionary tree using the character C as the root entry. The strings CA, CAP, CAPE, CAPER, CAPES, CAR, CART, CAT, CATS, CO, COM, COME, and CON are represented in this tree.

Procedures of V.42 bis

The V.42 bis Recommendation operates around five principal procedures (see Figure 7.9), which are described in the following paragraphs.

The string-matching operation matches a sequence of characters from the user data stream with an entry in the dictionary. For example, in the dictionary tree of Figure 7.8, a sequence of characters is matched against the dictionary entry of CAPER. The code word for CAPER is encoded into a number of bits, packed into an octet field, then passed to the control func-

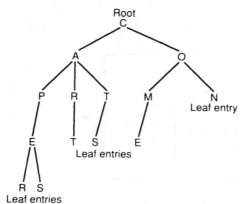

Figure 7.8 The V.42 bis tree.

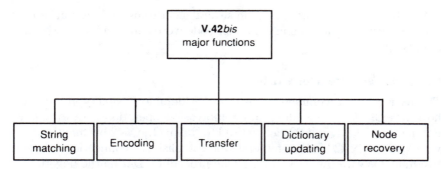

Figure 7.9 Major operations of V.42 bis.

tion. In turn, the control function passes this information to the error-control function, which encapsulates the compressed data into the LAPM I field.

The encoding operation is performed by the mapping of the matched code word of the dictionary into a binary value of N bits.

The transfer of the code words is the third major operation. The code words can be passed to the control function in either compressed or transparent mode.

The dictionary updating operation is used when a new entry to the dictionary is created. Let's assume that a character stream of CONE is matched against the dictionary. As shown in Figure 7.8, CONE does not exist in the compression tree. Therefore, only CON is passed to the encoding function for selection of an appropriate code word. The character E is added to the existing string of CON. A new code word is assigned to the leaf entry E of the tree CONE. Subsequent operations on CONE would result in the finding of a code word. As you will see shortly, the E is sent as part of the *next* code word and is used by the receiver to update its directory.

With the node recovery operation, an entry into the dictionary (a node) can be recovered for use in a subsequent dictionary update.

Before a dictionary is updated, V.42 bis requires that certain parameters be checked to make certain that code word sizes do not violate a threshold and that another code word can be added to the dictionary. The rules are governed by a number of parameters required by the data compression function:

- N1: Maximum size for the code word (in bits)
- N2: Total number of code words existing in the dictionary
- C2: The current code word size
- C3: The threshold for the size of the code word change
- N4: Number of characters in the alphabet, where $N4 = 2N^3$

V.42 bis uses these parameters in several algorithms to determine a new code word size, to determine if another code word can be added to the dictionary, etc.

Operating parameters for V.42 bis

The user of this recommendation has considerable flexibility in dictating how several of the encoding and compression operations are executed. Parameters P0, P1, and P2 can be used to "tailor" the V.42 bis functions. As with V.42, the V.42 bis Recommendation establishes a connection between two modems with the XID HDLC frame procedure. During the negotiation phase, the XID frame is used to establish values for the following operating parameters:

- *Parameter P0.* This parameter is used to signify if compression is to be used. It also stipulates the directions of the compression.

- *Parameter P1.* This parameter is used to negotiate a value for N2, which specifies the number of code words to be used during the operation. A default value of 512 is established for P1. This is its minimum value. It can be negotiated to larger values if needed.

- *Parameter P2.* This parameter contains a value for N7, which is the maximum string length permitted during the operation. This parameter also has a default value, 6, which is the minimum value. The range permitted for P2 is from 6 to 250.

These parameters, plus additional information, are carried in the DLC subfield (illustrated in Figure 7.4). The group ID identifies the frame as a private parameter set that is defined in ISO 8885 AD3. The group length field simply describes the entire length for the remainder of the DLC subfield; it does not include the group ID or the group length field itself.

Following these two control fields, V.42 bis currently describes four parameter values. Each parameter value contains a parameter ID, a parameter length, and a parameter value. The four parameter value fields contain the following information:

- Parameter value 1 is coded as V42.

- Parameter value 2 is coded as:
  ~Compression in either direction = 00
  ~Negotiation through initiator-responder direction only = 01
  ~Negotiation through responder-initiator direction only = 10
  ~Negotiation in either direction = 11

- Parameter value 3 contains the value of parameter P1.

- Parameter value 4 contains the value of parameter P2.

Example of the V.42 bis operations

This section describes in more detail the compression, decompression, and dictionary operations of V.42 bis. My thanks go to British Telecom (BT) for their assistance in explaining how BT operates its dictionary, which is not defined in V.42 bis.

Figure 7.10 shows a general view of an uncompressed and compressed data stream. In this example, two uncompressed fields, CONE and CONE-FLOWER, are sent as uncompressed images. The objective of V.42 bis is to create a code word for each of the character strings, CONE and CONE-FLOWER, and to send the code word instead. In this example, code word 1 (CW1) and code word 2 (CW2) are sent for CONE and CONEFLOWER.

Figure 7.11 shows the leaf entries for CAPE, COM, and CON. Assuming that the system is operating on the character string CONE, the next operation is to perform a lookup on this node to determine if CONE exists. Obviously, a quick look at Figure 7.11 reveals that CONE is not in the dictionary.

In Figure 7.12, the character string E has been added as a new leaf entry to node C. This addition has created a new concatenated character string of CONE. Shortly, I will return to this example to examine the results of the code word lookup and the dictionary.

Figure 7.13 depicts a general view of the operations of the transmit and receive modems. The "argument" character string is matched against the

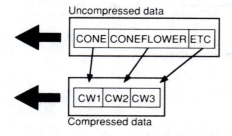

Figure 7.10 Compressed and uncompressed data.

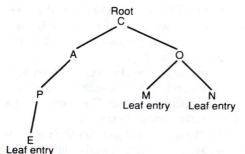

Figure 7.11 Leaf entries for CAPE, COM, and CON.

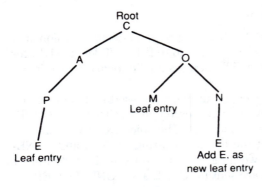

Figure 7.12 Addition of new entry to node C.

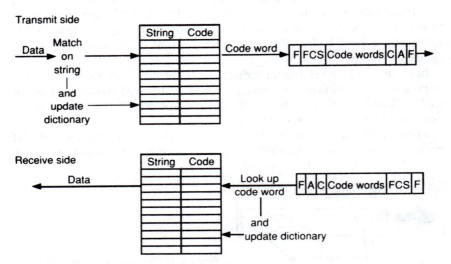

Figure 7.13 Transmitting and receiving dictionary.

directory on the transmit side, and an attempt is made to match this character string to a string existing in the dictionary. If a code word is found that matches (for a previous match, it could be CON but not CONE), the code word is retrieved for CON and sent to another buffer for assembly with other previously matched code words. The code words are placed in the LAPM frame and transmitted to the receiving modem.

The bottom of this figure shows the operations at the receiving modem. The code words are used to perform a lookup in the receiving modem's dictionary. The code word value reveals a particular string. This permits the code word to be mapped back to the original character string.

The dictionaries must be updated in both modems. Consider the transmit side first. When the dictionary detects the first character in the data stream that does not match an entry in the dictionary, it forwards the character string that did match with the appropriate code word, then uses the unmatched character to produce one more concatenated string in the dictionary. In this example, it sends the code word for CON and adds the CONE as a new string entry and with an appropriate code word.

It appears that the receive directory has an easy job of it, in that it simply finds the string that matches the code word for CON. However, it is important to understand that the transmit side is one step ahead of the receive side. Therefore, at the receiving modem, the first character of the next received code word, which will be E, must be used to update the receive side dictionary. The V.42 bis receiver always assumes that the first character of each string is to be used to update the previous character string. To repeat this important point, the transmit side is one step ahead of the receive side in the dictionary update cycle.

The conventional node tree, which was discussed earlier, is difficult to implement in programming logic. Therefore, the concept of the data structure tree is redrawn in what is known as a *TRIE*. This TRIE data structure makes it quite simple to perform the compression operations as well as the dictionary updates. The TRIE for the node C is depicted in Figure 7.14. Notice the arrows that point down or to the right. You are encouraged to compare Figure 7.14 with Figure 7.12 and to attempt to determine the hierarchical relationships of the characters in the figures. If you do not want to play games, read on and the relationships will be explained.

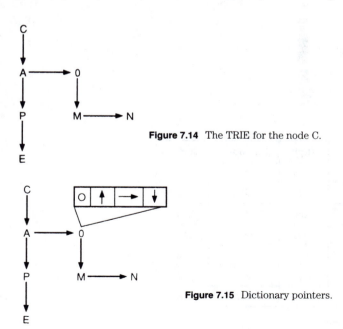

Figure 7.14 The TRIE for the node C.

Figure 7.15 Dictionary pointers.

Figure 7.15 explains the use of these arrows using the character O as the node example. The "blow up" at the top of the figure shows the node O with the three arrows next to it. The upper (↑) arrow is a symbol pointing to O's parent. This means that the character C is the parent of O. Next, the right (→) arrow points to the next entry in the TRIE, which is equivalent to O. Looking at the entry in Figure 7.14, there is no equivalent on this plane of the TRIE. Thus, the actual dictionary would be coded with a null value (indicating no arrow). The last arrow pointing down (↓) points to the child of node O. Two children are under O in the data structure tree; therefore, the first entry points to M, and M in turn points to its lateral entry, N. As you will see shortly, these arrows are implemented with values that act as pointers into specific entries of the dictionary.

Figure 7.16 shows the relationship of the TRIE to the entries in the dictionary. The dictionary entries are shown in the bottom part of the figure.

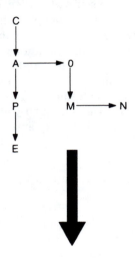

String	Parent	Right	Dependent	Code
C	Null	Null	A	312
A	C	O	P	313
P	A	Null	E	314
E	P	Null	Null	315
O	C	Null	M	316
M	O	N	Null	317
N	O	Null	Null	318

Figure 7.16 Dictionary entries.

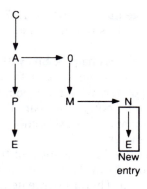

String	Parent	Right	Dependent	Code
C	Null	Null	A	312
A	C	O	P	313
P	A	Null	E	314
E	P	Null	Null	315
O	C	Null	M	316
M	O	N	Null	317
N	O	Null	E	318
E	N	Null	Null	319

Figure 7.17 A dictionary update.

Each entry consists of the character, the parent pointer, the next right pointer, the dependent pointer, and the code word that is transmitted in place of the character string.

Using Figure 7.16 as an example of a table lookup, if a match were to be made on the character string CAPE, the dictionary is entered by examining the string C, noting its dependent A; next, go to A and note A's dependent P; next, go to P and note P's dependent E; next go to E and end the matching operation. Therefore, the code word 315 would be transmitted for the character string CAPE.

This example does not go far enough because the dictionary lookup continues until a no match (NM) is found or until there is no more data. Therefore, let's move to Figure 7.17 to see the effect of adding a new entry by obtaining a no match in the dictionary. As before, the example for this exercise is CONE.

It can be seen from the entry in Figure 7.12 that CONE is added to the dictionary. From the standpoint of the TRIE in Figure 7.17, the N assumes a new child entry of E. Therefore, the update on entry N shows that the dependent null in Figure 7.16 was changed to E in Figure 7.17. The E entry then is added, its parent becomes N, and its rightmost and dependent entries are null.

In summary, the search through the dictionary begins with a character. An attempt is made to make a match on the next character by examining the right pointer or the dependent pointer. Whichever pointer leads to a match then leads to the next subsequent entry. The search will always go down the tree through the dependent pointer in the likelihood that the next entry will be a match with the right pointer.

In this example, the character string led to the entry of the string row to dependent A. Upon an examination of row A, the dependent P did not match, but the right node of O did match. Therefore, the next lookup goes into string O where a null match occurs with the right pointer, and there is no choice but to go to dependent M. For the letter M, you find a right pointer match on N. At this point, you have found CON. Proceeding down to N, you find there is a null at both right and dependent. Therefore, E becomes the dependent entry and is added to the dictionary. The additions are noted in the italicized areas in Figure 7.17. Compare them to Figure 7.16.

During this entry of CONE, remember that only the code word for CON was transmitted. When a no match occurs, the code word for CON (318) is transmitted and a search is *restarted* on node E.

Let's take the example one step further and assume that CONE has been successfully added to the dictionary. If the character string CONEFLOWER were to be transmitted, a match would occur through CONE and the code word 319 would be transmitted. When looking for the match for F, there would be a no entry from the C node. The code word 319 would be sent, and a search would be *restarted* at node F.

Figure 7.18 shows how the restart occurs. It shows the node entry TRIE for F. An examination reveals that F was stored with A, E, and L as children of F. Therefore, the match of CONEFLOWER would use the code word 319 for CONE and the code word 404 for FL. The O would be a no match (NM), so it would be added to the dictionary as code 405.

The bottom part of Figure 7.16 shows that the search of CONE provides code word 318. A no match (NM) then occurs. The FL match produces a code word of 404, and another no match occurs. Let's stop the example here; it's starting to become repetitive.

I have spent very little time with the operations at the receiver. The decoding operations are quite simple. The data structure is the same in the decoder as it is in the encoder. Decoding occurs by receiving a code word value, which is used as a lookup into the dictionary to locate the corresponding dictionary entry. The character is read and stored, and the parent pointer is used to locate the next entry (that is to say, the preceding entry). The decoding process follows a chain *back up through* the parent pointer

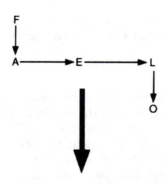

String	Parent	Right	Dependent	Code
F	NULL	NULL	A	401
A	F	E	NULL	402
E	F	L	NULL	403
L	F	NULL	O	404
O	C	NULL	M	405

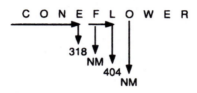

Figure 7.18 Restarting the search at another node.

to the initial character. The reading is in reverse order; however, the character string can be reconstructed by using a *push down stack* during the decoding and recovering the string afterward.

Finally, remember that the decoder operation must add the first character of its decoded string to the previous string.

Vendor-Specific Compression Schemes

Due to the slow development of V.34 and some dissatisfaction with V.42 bis, some vendors offer proprietary "V.Fast" modems and proprietary compression schemes. You are encouraged to carefully check the vendor's specifications. Also, many modems use MNP compression, and not V.42 bis. A common practice is to employ a 2:1 compression algorithm on a 9600 bit/s modem to achieve a rate of 19,200 bit/s.

Other vendors use proprietary schemes, with a combination of trellis coded modulation and V.42, to achieve even higher data rates. For example, the 32FAST modem from Black Box, Inc. can operate at 72 kbit/s on synchronous transmissions and at 115.2 kbit/s for asynchronous transmissions.

Summary

In the past, the ITU-T V Series Recommendations did not define any significant error-control measures beyond the physical layer. Recently, that trend has been changing, and with the publication of V.42 and V.42 bis, ITU-T has become an authoritative body for the definition of error-control measures, at both the physical and link levels. V.42 and V.42 bis have become the accepted standards for error-control and data-compression functions in modems. Today, practically all vendors use these standards in their product lines.

Transmission Quality and Maintenance

The V Series contains a number of recommendations dealing with transmission quality and maintenance. Some of these specifications are rather antiquated; others are brief.

Figure 8.1 depicts the organization of this section. The titles and identification numbers for the transmission quality and maintenance recommendations are:

- V.50: Standard limits for transmission quality of data transmission
- V.51: Organization of the maintenance of international telephone-type circuits used for data transmission
- V.52: Characteristics of distortion and error-rate measuring apparatus for data transmission
- V.53: Limits for the maintenance of telephone-type circuits used for data transmission
- V.54: Loop test devices for modems
- V.55: Specification for an impulsive noise-measuring instrument for telephone-type circuits
- V.56: Comparative tests of modems for use over telephone-type circuits
- V.57: Comprehensive data test set for high data signalling rates
- V.58: Management information model for V series DCEs

Transmission Quality and Maintenance

- V.50: Standard limits for transmission quality
- V.51: Maintenance of international telephone circuits
- V.52: Distortion and error-rate measuring apparatus
- V.53: Maintenance limits of telephone circuits
- V.54: Loop test devices
- V.55: Voice measuring
- V.56: Comparative tests for modems
- V.57: Data test set for high signalling rates
- V.58: Management information model for V Series DCEs

Figure 8.1 Transmission quality and maintenance recommendations.

V.50

This specification was published in 1968. It defines the transmission quality limits on telephone lines for data transmission. It establishes the limits on the degree of sending distortion and transmission channel distortion for V.21 and V.23 modems on leased and switched circuits. Because V.21 and V.23 modems are little used today, V.50 does not play a prominent role in a data communications system.

V.51

The recommendation defers to the ITU-T Recommendation M.729. The M.729 Recommendation is a brief description of a procedure to be established in each country for problem resolution. The coordinator for this activity is called the *data coordinating point* (DCP).

The purpose of the DCP is to act as a contact between telephone administrations to investigate transmission problems. The DCPs are required to agree on courses of remedial action in order to repair telephone transmission problems. M.729 makes no discussion of any international circuit arrangements.

M.729 also defers discussion to other M Recommendations for establishing certain procedures for problem resolution. These procedures are beyond the scope of this book. You might want to obtain Recommendations M.710, M.580, and M.605 for more information on this subject.

V.52

In the Red Book (1984), V.52 provided guidelines for measuring the distortion on telephone lines for data transmission. In the Blue Book, V.52 has been replaced with Recommendation O.153.

O.153 provides guidelines for measuring the error performance of asynchronous or synchronous data circuits that operate between 0.050 and 168 kbit/s. The recommendation describes the use of the following test patterns:

- 511-bit pseudo-random test pattern for data rates up to 14.4 kbit/s
- 2047-bit pseudo-random test pattern for data rates of 64 kbit/s
- 1048-bit pseudo-random test pattern for data rates up to 72 kbit/s
- Several fixed test patterns for continuity tests, such as alternating space/mark, "QUICK BROWN FOX," etc.

O.153 also specifies the clock sources and the interfaces that are to be used for the testing apparatus. For example, V.10, V.11, V.24, V.35, and V.36 of the V Series are specified. The specific interface chosen is a matter that is left to the user.

V.53

V.53 defines the parameters to indicate the quality of transmission on the channel. Again, it is an older specification that deals with V.21 and V.23 modems, with modulation rates from 200 to 1200 bauds, for both leased and switched lines. It establishes parameters for the bit error rate with the various modems, line connections, and baud based on the testing bit sequence of 511 bits.

V.54

V.54 has been more widely used than some of the other recommendations in this section. It defines the testing procedures for the connections between the data circuit-terminating equipment (DCEs) and data terminal equipment (DTEs) and between the DCEs themselves.

V.54 defines the following procedures for link testing. A problem communications link is tested by placing the modems in a loopback mode. Practically all modems can be put into loopback tests automatically, programmatically, or with a switch on the modem (see Figure 8.2). The loopback signals are analyzed to determine their quality and the bit error rate resulting from the tests. V.54 also defines how the V.24 interchange circuits are to be used during the four loopback operations.

Loop 1

The DTE interface logic can be tested with an *internal loopback* (loop 1). V.54 recommends that a loop 1 test be conducted as close to the interface

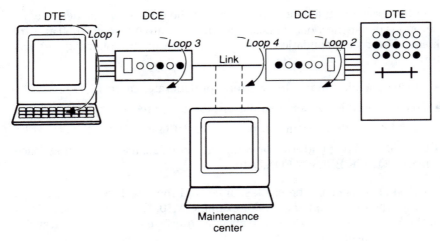

Figure 8.2 Loopback testing with V.54.

as possible. The specification defines the operations of several of the V.24 interchange circuits for the loop 1 test:

- Circuit 105 is in an off condition.

- Circuit 108/1 or 108/2 must be in the same condition that existed before the test was conducted.

- Circuit 103 must be connected to circuit 104 within the DTE.

- The DTE should continue to monitor circuit 125 to allow an incoming call to be given precedence over this test.

- Circuit 103 must be in a binary 1 condition.

Loop 3

The loopback can be sent through the local modem to test its analog and digital circuits; this test is called a *local loopback* (or loop 3). V.54 recommends that this loop be placed as close as possible to the communications link. In addition, this loop can be conducted on four-wire or two-wire circuits.

V.54 sets up the following requirements for a loop 3 test:

- Circuit 125 is to be monitored by the DTE to allow an incoming call to be given priority over this test.

- The data channel contains no traffic.

- The transmission line must be suitably terminated depending on the carriers or administration rules.

- The V.24 interchange circuits are to be operated normally, except for a two-wire half-duplex operation in which circuits 105 and 109 are disabled.

Loop 4

If the bit error rate is not beyond a specified level, the next step is likely to be a *remote line loopback* that tests the carrier signal and the analog circuitry of the remote modem (loop 4). The remote modem must be placed in the loopback mode for this test to be completed. V.54 establishes that loop 4 is used only on four-wire lines.

Loop 2

Loop 2 can be used only with a full-duplex DCE. V.54 has not yet defined the use of the backward channel for this test. V.54 establishes the following requirements for the use of the V.24 interchange circuits at the remote DCE:

- Circuits 103, 106, and 107 are in the off condition.
- Circuit 109 is connected internally to circuit 105.
- Circuit 104 is in a binary 1 condition.
- Circuit 103 is connected to circuit 104 inside the DCE.
- Circuit 115 is connected to circuit 113 inside the DCE (if these circuits are used).
- Circuits 115 and 114 must continue to operate with the DTE (if these circuits are provided).

Care should be taken in drawing conclusions from the remote analog loopback tests because the looped signal might be tested at twice its specification. For this reason, it often is advisable to send the signal through the remote modem digital circuitry to boost the signal power.

V.54 defines other types of testing. The testing of multipoint circuits, for example, is described both for synchronous and asynchronous DCEs. V.54 also specifies the configuration of the V.24 interchange circuits where, for example, 104 is clamped to 103, etc.

If you want detailed information on the operations of these loopbacks, Annex A of V.54 is highly recommended. It contains the state diagrams for the various testing procedures.

V.55

This recommendation refers to ITU-T Recommendation O.71. The O.71 Recommendation briefly (the document is two pages long) specifies the characteristics of a testing device to measure impulsive noise on telephone-type circuits. Operating parameters—such as attenuation and frequency values, bandwidth requirements, input impedance, etc.—are defined in O.71.

V.56

V.56 is a rather brief recommendation consisting mainly of tables that describe the types of distortions on the line, the test parameters for these distortions, and acceptable parameters within the distortions. As an example, V.56 defines attenuation distortion for the voice-band frequencies, delay distortion for these frequencies, etc.

V.57

V.57 has been replaced by Recommendation O.153, which was described earlier in this chapter.

V.58

As of this writing, V.58 has not yet been released by the ITU-T Telecommunications Management (TMN) object classes that are required for V Series network elements. You might want to study ITU-T Recommendation M.3010 for a description of TMN concepts and architecture.

Summary

The end user of the communication system usually is not aware of the V.50 Recommendations. They are used by PTT administrations and common carriers. The one exception to this statement might be V.54. Some of your company's technical representatives probably use this standard for loop-back testing and diagnostic work. With this exception, these important recommendations remain transparent to the end user.

Interworking with Other Networks

This part of the V Series Recommendations contains specifications on how certain signals are exchanged (and changed) between different types of networks. These standards have become increasingly important in the past few years as networks, carriers, and telephone administrations have implemented integrated services digital network (ISDN) based systems.

The Blue Book contains the following interworking specifications (see also Figure 9.1):

- V.100: Interconnection between public data networks (PDNs) and the public switched telephone network (PSTN)

- V.110: Support of data terminal equipments (DTEs) with V-Series type interfaces by an integrated services digital network (ISDN)

- V.120: Support by an ISDN of data terminal equipments (DTEs) with V-Series type interfaces with provision for statistical multiplexing

- V.230: General data communications interface layer 1 specification

V.100

The connection of user workstations to data networks through the public telephone network is rather common today. To ensure that the dial-and-answer telephone procedures are consistent across different manufacturers' equipment, the ITU-T has published V.100. This recommendation describes the procedures for physical-layer handshaking between answering and calling modems. The recommendation defines procedures for both half- and full-duplex procedures.

Interworking

- V.100: Interconnection of packet and telephone nets
- V.110: V series interfaces and ISDN
- V.120: V series, ISDN, and multiplexing
- V.230: General interface layer 1 specification

Figure 9.1 The V Series interworking recommendations.

V.100 requires that the dial-and-answer procedures of V.25 or V.25 bis be used to perform the initial handshaking between the modems. Among other requirements, the receiving modem must send a 2100-Hz answer tone back to the transmitting modem. Once it has transmitted this tone to the receiving modem, it enters the operations defined in V.100.

As depicted in Figure 9.2, once the modems have exchanged the dial-and-answer tones, the answering modem transmits what is known as the S1 signal. This signal is of a certain frequency, depending upon the type of mo-

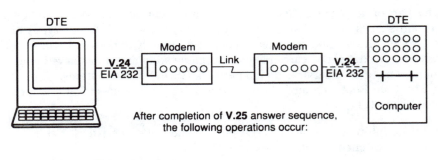

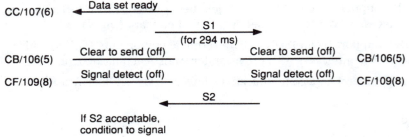

Where: XX/YYY(ZZ)

    XX = EIA-232-D interchange circuit designators
    YYY = V.24 interchange circuit designators
    (ZZ) = ISO 2110/EIA pin assignments

Figure 9.2 V.100 handshaking procedures.

dem used. As an example, if the answering modem is a V.22 bis modem, operating with duplex, synchronous procedures, the S1 signal is an 1800-plus 2250-Hz signal. As another example, if the answering modem is a V.23 modem, its S1 signal is a 1300-Hz tone.

Upon sending signal S1, the modem remains silent until it detects a signal S2. Based on its response to S2, it either disconnects or conditions itself to the selected mode as indicated in S2. To continue the previous example, if the modem sends an S1 signal indicating it is a V.22 bis modem, the other modem sends back an S2 signal of 1200 Hz. Using the second example, upon receiving the V.23 S1 signal from a modem, the other modem responds with a 390-Hz S2 signal.

V.100 provides a number of calling and called scenarios for the S1 and S2 signals. As you can see, the recommendation provides a very useful protocol for identifying unknown modem types.

V.110

The V.110 Recommendation has received considerable attention in the industry because it defines procedures that have been incorporated into several vendors' ISDN terminal adaptors (TAs). Among other features, V.110 establishes the conventions for adapting a V Series data rate to the ISDN 64 kbit/s rate. Figure 9.3 illustrates the scheme used by V.110.

The V.110 terminal adapter consists of two major functions: rate adapter 1 (RA1) and rate adapter 2 (RA2). The RA1 produces an intermediate rate (IR), which then is input into RA2. RA1 accepts standard V Series interface data rates, ranging from 600 bit/s to 19,200 bit/s.

Based on the data rate that is transmitted from the user device, RA1 produces one of the following intermediate rates: 8, 16, or 32 kbit/s. Bit rates of 48 and 56 kbit/s are not processed by the rate adapters; they are converted directly by the terminal adapter into a 64-kbit/s B channel rate.

Table 9.1 summarizes the intermediate rate produced by RA1.

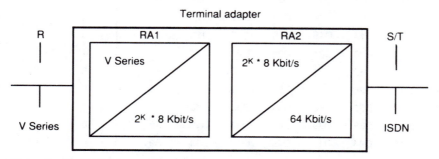

Figure 9.3 The V.110 bit rate adaptation scheme.

TABLE 9.1

V Series interface rates (bit/s)	Intermediate rate (RA1; kbit/s)
600	8
1200	8
2400	8
4800	8
7200	16
9600	16
12,000	32
14,400	32
19,200	32

As shown in Figure 9.3, the intermediate rate is calculated by the formula $2^k \times 8$ kbit/s where k can take the values of 0, 1, or 2.

Figure 9.4 illustrates the output of RA1; it is an 80-bit frame (the IR frame). The user data are placed into this frame. The bits in the frame also are used for a variety of control functions, which are examined in more detail shortly:

- Seventeen bits are used for synchronization to provide frame alignment patterns.

- Several bits are used to convey information about the status of V.24 circuits 108, 107, 105, 109, and 106.

- Several bits are used for network independent clocking information.

- Several bits are used in conjunction with the IR to provide the user data signalling rate for synchronous transmissions.

			Bit position					Octet position
1	2	3	4	5	6	7	8	
0	0	0	0	0	0	0	0	0
1	D1	D2	D3	D4	D5	D6	S1	1
1	D7	D8	D9	D10	D11	D12	X	2
1	D13	D14	D15	D16	D17	D18	S3	3
1	D19	D20	D21	D22	D23	D24	S4	4
1	E1	E2	E3	E4	E5	E6	E7	5
1	D25	D26	D27	D28	D29	D30	S6	6
1	D31	D32	D33	D34	D35	D36	X	7
1	D37	D38	D39	D40	D41	D42	S8	8
1	D43	D44	D45	D46	D47	D48	S9	9

Figure 9.4 The V.110 frame format.

Bit position								Octet
1	2	3	4	5	6	7	8	position
0	0	0	0	0	0	0	0	0
1	D1	D2	D3	D4	D5	D6	S1	1
1	D7	D8	D9	D10	D11	D12	X	2
1	D13	D14	D15	D16	D17	D18	S3	3
1	D19	D20	D21	D22	D23	D24	S4	4
1	E1	E2	E3	E4	E5	E6	E7	5
1	D25	D26	D27	D28	D29	D30	S6	6
1	D31	D32	D33	D34	D35	D36	X	7
1	D37	D38	D39	D40	D41	D42	S8	8
1	D43	D44	D45	D46	D47	D48	S9	9

Figure 9.5 The data bits.

As shown in Figure 9.4, each frame consists of 80 bits. The contents of the bits vary and depend on the data signalling rate from the user device as well as the use of the V.24 interchange circuits. The bit positions are arranged with the bits in an octet shown on the top of the figure. Octets are stacked below each other as noted in the right-hand side of the figure. The order of bit transmission is low-order bit first, with the first octet first.

Figure 9.5 shows the positions of the user data (the D bits) inside the V.110 frame. The bit positions are in italics to aid you in determining their location. A maximum of 48 bits can be sent in each frame. Therefore, up to 19,200 bits can be placed in the IR frame, which as you learned earlier, has a maximum rate of 32 kbit/s. This value can be derived from a simple calculation: A 32-kbit/s channel allows 400 frames to be transmitted per second (32,000 divided by 80 equals 400). A maximum of 48 bits can be placed in each frame; therefore, 400 times 48 equals 19,200. If a smaller data signalling rate is used (such as 600 bit/s), some of the positions in the frame are not relevant.

Figure 9.6 shows the positions of the synchronization bits (which are in italics). These bits are used to synchronize the machines' transmissions. As the figure shows, the first octet serves as the initial synchronization signal and is set to all 0s. Thereafter, bit 1 of each of the following 9 octets (set to 1) completes the synchronization pattern.

The S and X bits (italic in Figure 9.7) are used to provide the mapping functions of several of the V.24 interchange circuits that exist at the local user device (DTE) and the local TA (read data circuit-terminating equipment, or DCE). The state of these interchange circuits is mapped into the S and X bits, sent across the channel to the remote TA-DTE interface, then

			Bit position					Octet position
1	2	3	4	5	6	7	8	
0	*0*	*0*	*0*	*0*	*0*	*0*	*0*	0
1	D1	D2	D3	D4	D5	D6	S1	1
1	D7	D8	D9	D10	D11	D12	X	2
1	D13	D14	D15	D16	D17	D18	S3	3
1	D19	D20	D21	D22	D23	D24	S4	4
1	E1	E2	E3	E4	E5	E6	E7	5
1	D25	D26	D27	D28	D29	D30	S6	6
1	D31	D32	D33	D34	D35	D36	X	7
1	D37	D38	D39	D40	D41	D42	S8	8
1	D43	D44	D45	D46	D47	D48	S9	9

Figure 9.6 The synchronization bits.

			Bit position					Octet position
1	2	3	4	5	6	7	8	
0	0	0	0	0	0	0	0	0
1	D1	D2	D3	D4	D5	D6	*S1*	1
1	D7	D8	D9	D10	D11	D12	*X*	2
1	D13	D14	D15	D16	D17	D18	*S3*	3
1	D19	D20	D21	D22	D23	D24	*S4*	4
1	E1	E2	E3	E4	E5	E6	E7	5
1	D25	D26	D27	D28	D29	D30	*S6*	6
1	D31	D32	D33	D34	D35	D36	*X*	7
1	D37	D38	D39	D40	D41	D42	*S8*	8
1	D43	D44	D45	D46	D47	D48	*S9*	9

Figure 9.7 The status bits.

used to operate the V.24 interchange circuits. With this approach, the system signals with digital signals, and no modems are required for the traffic inside the ISDN.

The italic E bits shown in Figure 9.8 provide several functions. Bits E1, E2, and E3 are used to identify the IR that is being used in the frame. These bits are manipulated to indicate IRs of 8, 16, or 32 kbit/s. Within these IRs, the three bits are manipulated to indicate the actual data rates of the devices. The combinations of 600, 1200, 2400, 4800, 7200, 9600, 12,000, 14,400, and 19,200 rates can be coded into these three E bits.

Bits E4, E5, and E6 are optional and can be used to carry network clocking information. Such information might be required if signals are received from devices outside of an ISDN system. As an example, a modem on a public telephone network might not be synchronized to the ISDN. These bits can be used to develop phase measurements for signalling synchronization.

The last E bit is the E7 bit. It is used to provide compatibility with Recommendation X.30 (I.461).

Mapping the V Series rate to the intermediate rate and the ISDN basic rate

A reasonable question to ask is "How does the V Series rate map to the IR, and how does the IR map to the ISDN basic rate?" A few simple calculations will give you the answer. First, assume that the TA creates an IR of 8 kbit/s. Because the IR frame is 80 bits long, the TA operates the IR at 100 frames per second: 8000 / 80 = 100.

Given a 2400-bit/s modem, it is obvious that it will not need all of the D bits in the IR frame. Therefore, the TA maps the 2400 bits in 24 slots of each frame (and pads the other D positions with redundant data bits) to yield a transfer rate of 2400 bit/s for user data within the 8-kbit/s intermediate rate transfer rate. There are 24 slots in each 80-bit frame (the other 24 bits are not needed), and the rate is 100 frames per second; therefore, $24 \times 100 = 2400$ bit/s.

All data and control bits in the IR frame must be sent across the ISDN to the remote user device, which means that all of the bits in the 80-bit frame

Bit position								Octet position
1	2	3	4	5	6	7	8	
0	0	0	0	0	0	0	0	0
1	D1	D2	D3	D4	D5	D6	S1	1
1	D7	D8	D9	D10	D11	D12	X	2
1	D13	D14	D15	D16	D17	D18	S3	3
1	D19	D20	D21	D22	D23	D24	S4	4
1	E1	E2	E3	E4	E5	E6	E7	5
1	D25	D26	D27	D28	D29	D30	S6	6
1	D31	D32	D33	D34	D35	D36	X	7
1	D37	D38	D39	D40	D41	D42	S8	8
1	D43	D44	D45	D46	D47	D48	S9	9

Figure 9.8 The E bits.

must be transferred. An ISDN basic rate provides for two B channels and one D channel. It processes each ISDN frame as follows:

- Each basic rate frame is 250 ms long.
- Therefore, 4000 frames are transferred per second: 1 s / 0.000250 = 4000.
- Each basic rate frame is 48 bits long, with 12 bits used to manage the basic rate channel, yielding a transfer rate for user data and control bits of 192 kbit/s: $4000 \times 48 = 192,000$.
- Therefore, the user data transfer rate is 144 kbit/s: $4000 \times (48 - 12) = 144,000$ (which equates to the 2B + D rate of $64 + 64 + 16 = 144$).

ISDN requires that an 8-kbit/s IR frame be multiplexed in the B slots within the basic rate frame as follows:

- The 8-kbit/s bits occupy bit position 1 of each B slot.
- Two B slots are allotted for each B channel in each 48-bit basic rate frame.
- Therefore, the 8-kbit/s IR occupies 2 bits in each basic rate frame.
- Because there are 4000 frames per second on the basic rate channel, using bit position 1 in the two slots of one of the B channels achieves the required intermediate rate of 8 kbit/s: $2 \times 4000 = 8000$.

Similar calculations can be performed for the other V.110 IRs, but this one exercise should give you an understanding of how the rates are mapped. The basic rate also could be mapped into the primary rate, which in turn could be mapped into yet higher rates (for example, broadband ISDN).

V.110 handshaking

Figure 9.9 illustrates one possibility in the V.110 handshaking protocol. The approach is quite simple. The TA sends 1s in the B and D channels and to the DTE until the DTE turns on V.24 108.2 (EIA-232-E CD). Then, the TA sends an 80-bit frame in the B channel with the S and X bits set to off. When data are sent across V.24 103 (EIA-232-E BA), the 80-bit frame's S and X bits are used to condition the remote V.24 interchange circuits.

V.110 asynchronous interfaces

The V.110 Recommendation also defines procedures for the adaptation of asynchronous rates as well as synchronous rates. To this point in the discussion, I have been dealing with only synchronous rates. Figure 9.10 shows the TA for the asynchronous rate adaptation scheme. It contains one additional function that is labeled "stop bit manipulation." This function is the RA0, which sits in front of RA1 and RA2 functions. It basically is an asynchronous-to-synchronous converter.

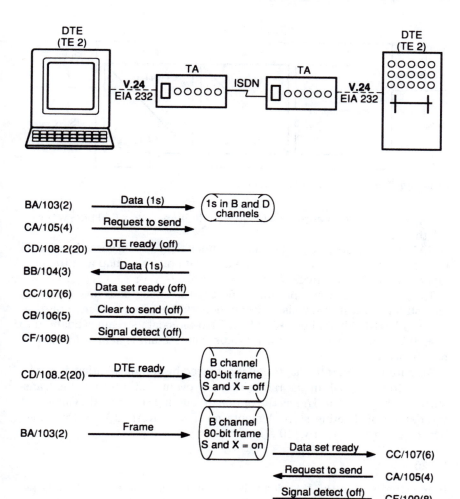

Where: XX/YYY(ZZ)

    XX   = EIA-232-D interchange circuit designators
    YYY = V.24 interchange circuit designators
    (ZZ) = ISO 2110/EIA pin assignments

Figure 9.9 V.110 handshaking.

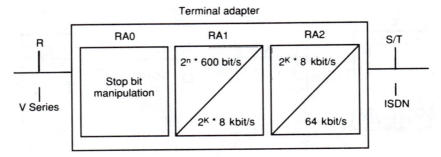

Figure 9.10 Adaptation for asynchronous rates.

RA0 produces a synchronous bit stream defined as $2^n \times 600$ bits, where n = 0 through 5.

RA1 and RA2 operate in the same manner that was discussed earlier in this chapter. The RA1 will accept the user data rate and map it to the next higher rate expressed by $2^k \times 8$ bit/s, where k = 0, 1, or 2.

The RA0 function accepts incoming asynchronous data and pads this data stream with stop signals to accommodate the nearest channel rate defined by $2^n \times 600$ bit/s. For example, a 7200-bit/s data stream is adapted to a 9600-bit/s data stream, and a 110-bit/s data stream is adapted to a 600-bit/s data stream.

As stated earlier, rate adaptation is achieved by adding additional stop elements to the signal. In the rather unlikely event that a terminal is transmitted overspeed, the TA has the option of deleting stop signals. Obviously, the degree of deleting stop elements must be governed by a threshold, which also is defined in V.110.

V.120

In recognition that the V Series will be in existence for a considerable period and in view of the need for ISDN devices (TE1) to interwork with non-ISDN devices (TE2), the ITU-T has published the V.120 Recommendation. V.120 supports an ISDN interface with a DTE and its associated physical-layer interface. The DTE is assumed to be operating with the V Series interfaces. V.120 also supports the multiplexing of multiple user data links onto the ISDN S/T interface. V.120 uses a link-level protocol based on the modification of LAPD (Q.921).

V.120 describes the use of a TA for the ISDN-to-V Series DTE interworking. However, this TA performs more functions than the TA we examined with V.110. It must perform the following:

- Electrical and mechanical interfaces conversions
- Adaptation of bit transfer rate (as in V.110)

- End-to-end synchronization of traffic
- Call management between the two end users
- A variety of maintenance functions

The scheme for V.120 is shown in Figure 9.11. Notice that the ISDN network sits between the TAs and TEs. V.120 establishes the protocols that are to be used between the TAs. The term TA-V means the TA supports a V Series interface.

Three modes of operations are supported with the V.120 terminal adapter: asynchronous, synchronous, and transparent. You probably are aware that asynchronous mode terminals (async TE2s) use start/stop bits and parity checks. The TA accepts the asynchronous stream from the user device and removes the start/stop bits. As an option, parity can be checked by the TA. In either case, the user data characters are placed in a frame for transmission to a peer entity. The peer entity is another TA or a TE1.

With the synchronous mode, the transmission from user TE2 is a high-level data link control (HDLC) type of frame. The HDLC flags and any zero-stuffed bits are removed at the TA. The TA performs an error check using the frame check sequence (FCS). In the event that an FCS check indicates that the frame was damaged during transmission, an FCS error information signal is relayed to the peer entity. It has the option of discarding the frame, generating an incorrect FCS for handling at the other side, or causing an abort at its interface (at the R-reference point).

The user device's address and control and information fields must flow transparently across the TA. The responsibility of the TA is to place these fields in a modified LAPD frame for transmission to its peer entity. The syn-

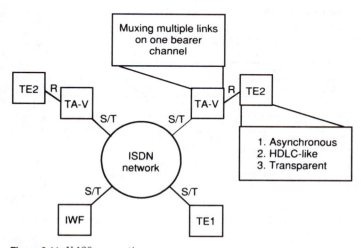

Figure 9.11 V.120 connections.

chronous mode also allows the segmentation of a user message into smaller pieces. The TA can perform this multiplexing operation to avoid the delay involved in waiting for the complete bit stream of the message. The receiving peer entity is required to perform a demultiplexing process at the other end.

The TA also can support bit-transparent operations. This option requires the TA to encapsulate the bits from the user device at the R-reference point. However, no processing of the bits occurs, nor is any error checking made. The bits are relayed transparently to the peer entity. The transparent mode of operation is to be used if the asynchronous and synchronous modes are not used. The transparent mode of operation can use the HDLC unnumbered information (UI) frame option.

V.120 message format

V.120 uses the frame structure of the Q.921/I.441 Recommendations. This format is illustrated in Figure 9.12. The fields relating to the HDLC frame structure are not discussed here. I will concentrate on the fields unique to this recommendation.

The logical link indicator (LLI) is a concatenation of the LLI0 and LLI1 fields. This 13-bit field can take the values of 0 to 9181. The LLI values are similar to the service access point identifier (SAPI) and the terminal end point identifier (TEI) fields that reside in the LAPD frame. Presently, the following values are reserved:

0	End channel signalling
1–255	Reserved
256	Default LLI
257–2047	For LLI assignment
2048–8190	Reserved
8191	In-channel layer management

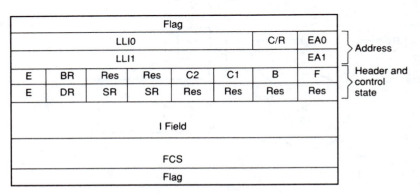

Figure 9.12 The V.120 frame format.

The extended address fields (EA) are used to extend the number of octets in the address. If the value of this bit is zero, it means that additional address octets follow. The presence of a 1 value in bit 1 of the address field means that it is the last octet of the address field. As with LAPD, the command/response (C/R) bit indicates a command (C = 0) or response (R = 1).

The V.120 header resides in front of the information (I) field as part of the HDLC I field. It consists of two parts: the header octet and the control state information octet.

The header octet has eight fields, each 1 bit long. The extension bit (E bit) is used to signify that the header has additional state information. If this bit is set to zero, it indicates that the control state information follows. Obviously, the control state information octet is optional.

The break bit (BR) is used in asynchronous applications to accommodate a terminal creating a break signal on the line. If the TA receives a break signal, it changes the BR bit to indicate the break. It must do this in the same frame or after the queued characters have been set. The asynchronous break signal is mapped by the TA to an HDLC idle condition. If the user device is an HDLC station, the TA sets this bit to 1 if it receives an idle signal from the HDLC station.

The next two bits are reserved and are set to zero. The following two bits are designated as C1 and C2, respectively. These bits are used for diagnostic purposes to notify the TA entities of various types of error conditions that have been detected. The combination of the two bits and their meaning is dependent upon the use of the synchronous, asynchronous, or bit-transparent modes.

The B and F bits (beginning and final) are used to convey information about the segmentation of the message. These bits are used primarily for synchronous mode operations. The values for synchronous modes are as follows:

10 Begin frame
00 Middle frame
01 Final frame
11 Single frame

The value mode 11 also pertains to asynchronous and bit-transparent modes of operations.

The second octet of the V.120 header is the control state information octet. It is optional and, as you learned earlier, designated with coding of the E-extension bit with a value of zero.

Presently, only four bits of this octet are designated for use. They are used as follows:

- Extension bit (E): Used to indicate a further extension of the header octet.

- Data ready bit (DR): Used to indicate that the TE1 interface is activated. At the transmitting end the DR bit is mapped from the V.24 circuit 108/2 data terminal ready (EIA CD). At the receiving end, no mapping is required for the DR bit.

■ Send ready (SR): Used to indicate that the TE is ready to send data. The SR bit is mapped from the V.24 105 interchange circuit request to send (EIA CA) and the V.24 109 interchange circuit receive line signal detect (EIA CF). At the sending end, this variable is mapped from circuit 105. At the receiving end, this bit is mapped to circuit 109.

■ Receive ready (RR): Used to indicate that the TE is ready to receive data. The RR bit is not mapped at the sending end. At the receiving end, the RR bit maps to the V.24 interchange circuit 106 ready for sending (EIA CB).

■ RES: These bits are reserved for future use.

V.120 logical connections

The V.120 Recommendation is closely aligned with Q.921 with regard to information transfer and acknowledgment of frames. There are several differences, which are beyond the scope of this book. If you want more information, you should refer to Section 2.4 of V.120, Section 5.2 of Q.921, and Section 5.5 of Q.921.

V.120 operations with asynchronous stations

Figure 9.13 shows an example of TA: supporting asynchronous data flow from a TE2 on the R reference point and supporting the flow of the frame from the ISDN S/T reference point. [The notation (O) means that the operation is optional.]

The traffic from a TE2 at the R interface is received at the TA, and as mentioned earlier, the TA removes the start/stop bits and performs a parity check. If the code is 8 bits long, it removes the parity. If the code is fewer than 8 bits, it is responsible for padding so that the character equals 8 bits.

As another possibility, a break signal that is transmitted from the user DTE is used to set the BR bit accordingly. Notice that the C1 and C2 bits also are set. In the event that a parity bit error is detected, the TA uses the C1 and C2 bits to indicate the following possibilities.

The TA sets the C1 bit to 1 upon detecting a parity error from the TE2 device. The character that was detected in error must be the last character in the frame that was transmitted. Consequently, the frame with the C1 set to 1 simply indicates that the last character in the frame has a parity error.

If a stop character is detected with a parity error, the TA sets C2 to 1. This action indicates that the last character in this frame is a stop bit and contains an error.

As you might expect, the TA is responsible for performing a complementary conversion function for the TE2 terminal to receive the traffic. As the

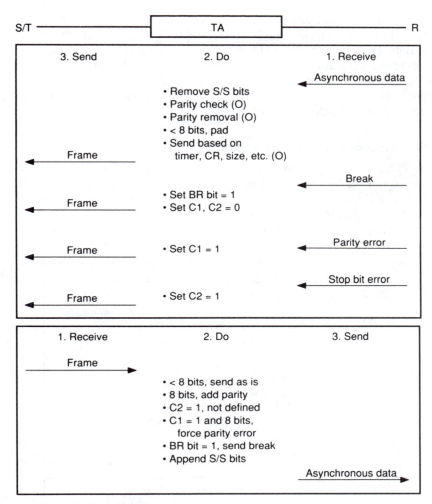

S/T ——————————— | TA | ——————————— R

3. Send   2. Do   1. Receive

Asynchronous data

• Remove S/S bits
• Parity check (O)
• Parity removal (O)
• < 8 bits, pad
• Send based on

Frame   timer, CR, size, etc. (O)

Break

Frame   • Set BR bit = 1
               • Set C1, C2 = 0

Frame   • Set C1 = 1   Parity error

Stop bit error

Frame   • Set C2 = 1

1. Receive   2. Do   3. Send

Frame

• < 8 bits, send as is
• 8 bits, add parity
• C2 = 1, not defined
• C1 = 1 and 8 bits,
    force parity error
• BR bit = 1, send break
• Append S/S bits

Asynchronous data

Figure 9.13 V.120 and asynchronous operations.

frames are received from the interface at the S/T reference point, the TA adds the start/stop bits and sends the traffic to the TE2. These operations are shown in the lower box of Figure 9.13.

If the TA receives a frame with the BR bit set to 1, it will send a break signal to the TE2. If the characters that were received are fewer than 8 bits, the TA forwards the bits as they are received. If the characters contain 8 bits, the TA adds the appropriate parity bits and sends the data to the TE2.

The TA handles the parity bit errors as follows. If the frame received has the C2 bit set to 1, ITU-T does not define the actions for the TA. However, if the C1 bit is set to 1 and the character code is an 8-bit code, the TA can force a parity error on the last character sent to the TE2. In other words, it intentionally corrupts a character.

Figure 9.13 is self-explanatory, but the operations that are depicted in the figure have been summarized in this discussion. You might want to go through the operations in the figure to gain a better understanding of how V.120 supports asynchronous interfaces.

V.120 operations with HDLC-like stations

If the TE2 is operating in synchronous mode and using an HDLC frame, the TA is responsible for removing beginning flags and any stuffed zeros. As illustrated in Figure 9.14, the TA also performs an FCS check to determine whether the data are corrupted on the link. The TA also removes the FCS characters as well as the ending flag. If necessary, the data are segmented

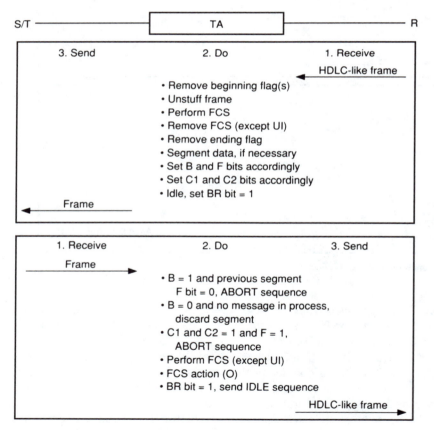

Figure 9.14 V.120 operations with HDLC-like stations.

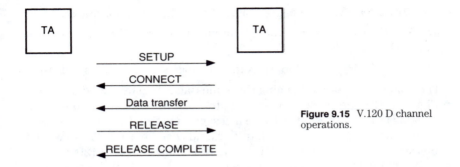

Figure 9.15 V.120 D channel operations.

and each segment is preceded by the V.120 header. The segmentation is defined by the use of the B and F bits, which were discussed earlier. The errors detected by the TA on a frame are coded by a combination of the C1 and C2 bits, also discussed earlier.

The traffic emanating from the ISDN across the S/T interface is handled by the TA in the following manner (see the lower box in Figure 9.14). The TA (re)calculates the FCS, unless the system is using the HDLC UI frame option. Additionally, it uses the V.120 header to determine the beginning segments with the B bit. It uses the C1 and C2 bits to determine any errors.

V.120 bit-transparent mode operations

The operations for bit-transparent mode are quite simple. The TA segments the incoming data stream from the TE2 and sends it to the S/T interface in the V.120 frames. It passes the bits coming from the S/T interface by simply removing the bits from the frame and passing them to the R interface.

Establishment of logical links within a bearer channel

The preceding information described how the TAs and TEs communicate at the physical and link layers. V.120 also provides a means for establishing logical connections (logical links) at the network layer. V.120 uses Q.931 procedures on the D channel for the establishment of these connections. The operation is shown in Figure 9.15.

The following messages are used for the establishment of logical links within the bearer channel:

- *SETUP.* Sent by either TA to initiate a new logical link. It uses a protocol discriminator value of 00001001 to distinguish it from the full Q.931 Recommendation.

- *CONNECT.* Sent by an accepting TA to indicate that a request for a logical link has been accepted.

- *Data Transfer.* Transfer of user data between the communicating devices.

- *RELEASE.* Indicates that the TA intends to release the call reference and the logical link.

- *RELEASE COMPETE.* Indicates that the logical link has been released.

The connection for exchanging these messages is established as shown in Figure 9.16. The protocol discriminator, call reference, and message-type fields must be present in these messages. Optionally, the lower layer capability field and logical link identifier fields can be used. These will be discussed shortly. The protocol discriminator value for the V.120 must always be 00001001. The call referenced values depend on the particular setup, and the message types reflect the Q.931 messages that were just listed.

The use of the lower layer compatibility field can be quite valuable because it allows the TAs to determine what type of lower layer interfaces are required. The contents of this field vary but may include the following:

- Information transfer rate

- Only UI frames supported

- Synchronous or asynchronous operations

- Bit or protocol sensitive mode

- Number of stop bits

- Number of data bits

- Parity

- Duplex mode

- Modem type

- Layer 2 and 3 protocols

Finally, the last field that can be used in the V.120 identifies the logical link. You should remember that the purpose of the LLI is to identify each logical link within the bearer channel. This value is carried in octets 3 and 4 of the LLI field. The logical link is established by either TA transmitting the SETUP message. The LLI value in the SETUP message is used to map the logical link onto the bearer channel. It is not necessary that the TA

Length	
1	Protocol discriminator
2	Call reference
1	Message type
2-13	Low layer compatibility
4	Logical link identifier

Figure 9.16 The Q.931 message.

assign the LLI value. If this is the case, the receiving TA will be given the responsibility of assigning this value by including the LLI field in the CONNect message.

V.230

Even though the ITU-T has fostered the ISDN technology during the last several years, the organization has recognized the need to provide interfaces between V Series equipment and the I Series Recommendations. Previous discussions in this section focused on several standards that provide these interfaces. Recommendation V.230 is yet another specification to ease the internetworking task between V Series and ISDN systems. V.230 addresses the physical layer signalling, synchronization, timing, and wiring to achieve this interworking.

V.230 describes the interfaces between DTE and DTEs, DCE and DCEs, and DTE and DCEs. It is based on ISDN Recommendation I.430. A separate specification is published to allow the use of different wiring configurations and to provide a simple protocol to allow the equipment to signal if it is using a V.230 or I.430 interface. Fortunately, the specifications contained in V.230 permit a terminal design that can be compatible with both I.430 and V.230.

The V.230 frame format

The formats for the frames exchanged between the TE and the network (NT) are shown in Figure 9.17. The formats vary in each direction of transfer but are identical for point-to-point or multipoint configurations. The frames are 48 bits long and are transmitted by the TE and NT every 250 μs. The first bit of the frame transmitted to the NT is delayed by two bit periods with respect to the first bit of the frame received from the NT.

The 250-μs frame provides 4000 frames a second (1 / 0.000250 = 4000) and a transfer rate of 192 kbit/s (4000 × 48 = 192,000). However, 12 bits in each frame are overhead, so the user data transfer rate is 144 kbit/s [4000 × (48 − 12) = 144,000].

The first two bits of the frame are the framing bit (F) and the dc balancing bit (L). These bits are used for frame synchronization. In addition, the L bit is used in the NT frame to electrically balance the frame and, in the TE frame, to electrically balance each B channel octet and each D channel bit. The auxiliary framing bit (F_a) and the N bit also are used in the frame alignment procedures. Bit A is used for TE activation and deactivation. The B1 and B2 bits contain the user data for the B channels.

Contending for use of the D channel

The user devices (TE1 or TE2 plus TA) can be multidropped onto one ISDN circuit. With this configuration, it is possible that the terminals might trans-

Master to slave

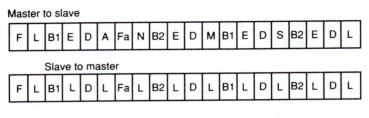

Slave to master

Where:

F = framing bit   B1 = 8 bits in B1 channel
L = DC balancing bit   B2 = 8 bits in B2 channel
D = DV channel bit   A = activation bit
E = DV echo-channel bit   S = S channel multiframe bit
Fa = auxiliary framing bit   M = Multiframing bit
N = inverted value of Fa

Figure 9.17 The V.230 frame format.

mit at approximately the same time, which would result in collisions. ISDN provides several features to determine if other devices are using the link at the same time.

A TE (called a slave in V.230) transmits in the D channel when it has determined that the channel is free. When the NT (called a master in V.230) receives a D bit from the TE, it echoes back the bit in the next E bit position. The TE expects the next E bit to be the same as its last transmitted D bit. Under normal conditions, the TE continues to detect its own D bits in the E bits.

A terminal cannot transmit into the D channel toward the NT until after it has detected at least eight consecutive 1s (no signal) corresponding to a preestablished priority. If the number is reached, the TE then can send its D channel data.

If the TE detects a bit in the echo channel (E channel) that is different from its D bits, it must stop transmitting immediately. This indicates that another terminal has begun transmitting at the same time.

This simple technique ensures that only one terminal can transmit its D message at one time. After the successful transmission of the D message, the terminal has its priority reduced by requiring it to detect more continuous 1s before transmitting. A terminal is not allowed to raise its priority back to its previous value until all other devices on the multidrop line have had an opportunity to send a D message.

Priorities can be set when the equipment is manufactured or it can be established by an upper layer. A telephone connection on the line usually has a higher priority and precedence over data services. Signalling information is a higher priority than nonsignalling information. Presently, the TE must detect 10 E bits of 1s before sending nonsignalling information but only 8 E bits before sending signalling information.

Summary

With the increased use of ISDN, the ITU-T has recognized the need to publish ISDN internetworking recommendations. This chapter described the four recommendations that describe the internetworking of V Series machines with ISDN networks. These standards are used extensively in the industry. You will most likely recognize some of these operations from reading a vendor's marketing brochure.

Framework for Link Level Protocols

A prerequisite to any discussion of the data link control layers is an understanding of the high-level data link control (HDLC) specification. It forms the basis for several of ITU-T's link-layer specifications such as link access procedure balanced (LAPB), link access procedure on the D channel (LAPD), and link access procedure for modems (LAPM).

HDLC is a link protocol specification that was published by the International Standards Organization (ISO) as ISO 3309 and ISO 4335 (and supporting documents 7809, 8471, and 8885). It has achieved wide use throughout the world. The recommended standard provides for many functions and covers a wide range of applications. It frequently is used as a foundation for other protocols which use specific options in the HDLC repertoire.

This appendix addresses the main functions of HDLC. The reader is encouraged to check with specific vendors for their actual implementation of HDLC. Most vendors have a version of HDLC available, although the protocol often is renamed by the vendor or designated by different initials.

HDLC Characteristics

HDLC provides for a number of link options to satisfy a wide variety of user requirements. It supports both half-duplex and full-duplex transmission, point-to-point and multipoint configuration, as well as switched or nonswitched channels. An HDLC station is classified as one of three types: primary, secondary, or combined.

The *primary* station is in control of the data link. This station acts as a master and transmits *command* frames to the secondary stations on the channel. In turn, it receives *response* frames from those stations. If the link is multipoint, the primary station is responsible for maintaining a separate session with each station attached to the link.

The *secondary* station acts as a slave to the primary station. It responds to the commands from the primary station in the form of responses. It maintains only one session, with the primary station, and has no responsibility for control on the link. Secondary stations cannot communicate directly with each other; they first must transfer their frames to the primary station.

The *combined* station transmits both commands and responses and receives both commands and responses from another combined station. It maintains a session with the other combined station.

HDLC provides three ways to configure the channel for primary, secondary, and combined station use: unbalanced, symmetrical, and balanced.

An *unbalanced* configuration provides for one primary station and one or more secondary stations to operate as point-to-point or multipoint, half-duplex, full-duplex, switched or nonswitched. The configuration is called unbalanced because the primary station is responsible for controlling each secondary station and for establishing and maintaining the link.

The *symmetrical* configuration is seldom used today. The configuration provides for two independent, point-to-point, unbalanced station configurations. Each station has a primary and secondary status. Therefore, each station is considered logically to be two stations: a primary and a secondary station. The primary station transmits commands to the secondary station at the other end of the channel and vice versa. Even though the stations have both primary and secondary capabilities, the actual commands and responses are multiplexed onto one physical channel.

A *balanced* configuration consists of two combined stations connected point-to-point only, half-duplex or full-duplex, switched or nonswitched. The combined stations have equal status on the channel and can send unsolicited frames to each other. Each station has equal responsibility for link control. Typically, a station uses a command to solicit a response from the other station. The other station can send its own command as well.

The terms *unbalanced* and *balanced* have nothing to do with the electrical characteristics of the circuit. Data link controls should not be aware of the physical circuit attributes. The two terms are used in a completely different context at the physical and link levels.

While the HDLC stations are transferring data, they communicate in one of the three modes of operation: normal response mode, asynchronous response mode, and asynchronous balanced mode.

Normal response mode (NRM) requires the secondary station to receive explicit permission from the primary station before transmitting. After receiving permission, the secondary station initiates a response transmission, which can contain data. The transmission can consist of one or more frames while the channel is being used by the secondary station. After the last frame transmission, the secondary station must wait for explicit permission before it can transmit again.

Asynchronous response mode (ARM) allows a secondary station to initiate transmission without receiving explicit permission from the primary station. The transmission can contain data frames or control information reflecting status changes of the secondary station. ARM can decrease overhead because the secondary station does not need a poll sequence to send data. A secondary station operating in ARM can transmit only when it detects an idle channel state for a two-way alternate (half-duplex) data flow or at any time for a two-way simultaneous (duplex) data flow. The primary station maintains responsibility for tasks such as error recovery, link setup, and link disconnections.

Asynchronous balanced mode (ABM) uses combined stations. The combined station can initiate transmissions without receiving prior permission from the other combined station.

NRM frequently is used on multipoint lines. The primary station controls the link by issuing polls to the attached stations (usually terminals, personal computers, and cluster controllers). The ABM is a better choice on point-to-point links because it incurs no overhead and delay in polling. The ARM is seldom used today.

The term *asynchronous* has nothing to do with the format of the data and the physical interface of the stations. It is used to indicate that the stations need not receive a preliminary signal from another station before sending traffic. HDLC uses synchronous formats in its frames.

Frame format

HDLC uses the term *frame* to indicate the independent entity of data (protocol data unit) transmitted across the link from one station to another. Figure A.1 shows the frame format. The frame consists of four or five fields:

Flag fields (F)	8 bits
Address field (A)	8 or multiple of 8 bits
Control field (C)	8 or multiple of 8 bits
Information field (I)	Variable length; not used in some frames
Frame check sequence field (FCS)	16 or 32 bits

All frames must start and end with the flag (F) sequence fields. The stations attached to the data link are required to continuously monitor the link for the flag sequence. The flag sequence consists of 01111110. Flags continuously are transmitted on the link between frames to keep the link in an active condition.

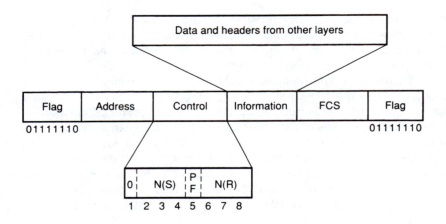

Flag: Delineates beginning and ending of frame
Address: Identifies link station
Control: Used for sequencing, flow control
Information: User data or control headers
FCS: Frame check sequence (for error checking)
N(S): Sending sequence number
N(R): Receiving sequence number
P/F: The poll or final bit

Figure A.1 The link layer frame.

Other bit sequences also are used. At least seven but less than fifteen continuous 1s (abort signal) indicate a problem on the link. Fifteen or more 1s keep the channel in an idle condition. One use of the idle state is in support of a half-duplex session. A station can detect the idle pattern and reverse the direction of the transmission.

Once the receiving station detects a nonflag sequence, it is aware it has encountered the beginning of the frame, an abort condition, or an idle channel condition. Upon encountering the next flag sequence, the station recognizes it has found the full frame. In summary, the link recognizes the following bit sequences as:

01111110   = Flags
At least 7 but less than 15 1s = Abort
15 or more 1s   = Idle

The time between the actual transmission of the frames on the channel is called *interframe time fill*. This time fill is accomplished by transmitting continuous flags between the frames. The flags can be 8-bit multiples and can combine the ending 0 of the preceding flag with the starting 0 of the next flag.

HDLC is a code-transparent protocol. It does not rely on a specific code (ASCII/IA5, EBCDIC, etc.) for the interpretation of line control. For example, bit position n within an octet has a specific meaning, regardless of the other bits in the octet. On occasion, a flaglike field, 01111110, might be inserted into the user data stream (I field) by the application process. More frequently, the bit patterns in the other fields appear "flaglike." To prevent "phony" flags from being inserted into the frame, the transmitter inserts a 0 bit after it encounters five continuous 1s anywhere between the opening and closing flag of the frame. Consequently, zero insertion applies to the address, control, information, and FCS fields. This technique is called *bit stuffing*. As the frame is stuffed, it is transmitted across the link to the receiver.

The procedure to recover the frame of the receiver is a bit more involved (no pun intended). The "framing" receiver logic can be summarized as follows. The receiver continuously monitors the bit stream. After it receives a 0 bit with five continuous, succeeding 1 bits, it inspects the next bit. If it is a 0 bit, it pulls this bit out; in other words, it unstuffs the bit. However, if the seventh bit is a 1, the receiver inspects the eighth bit. If it is a 0, it recognizes that a flag sequence of 01111110 has been received. If it is a 1, it knows an abort or idle signal has been received and counts the number of succeeding 1 bits to take appropriate action.

In this manner, HDLC achieves code and data transparency. The protocol is not concerned about any particular bit code inside the data stream. Its main concern is to keep the flags unique.

Many systems use bit stuffing and the nonreturn-to-zero-inverted (NRZI) encoding technique to keep the receiver clock synchronized. With NRZI, binary 1s do not cause a line transition, but binary 0s do cause a change. It might appear that a long sequence of 1s could present synchronization problems because the receiver clock would not receive the line transitions necessary for the clock adjustment. However, bit stuffing ensures that a 0 bit exists in the data stream at least every 5 bits. The receiver can use them for clock alignment.

The address (A) field identifies the primary or secondary station involved in the frame transmission or reception. A unique address is associated with each station. In an unbalanced configuration, the address field in both commands and responses contains the address of the secondary station. In balanced configurations, a command frame contains the destination station address and the response frame contains the sending station address.

The control (C) field contains the commands, responses, and sequence numbers that are used to maintain the data flow accountability of the link between the primary stations. The format and the contents of the control field vary, depending on the use of the HDLC frame.

The Information (I) field contains the actual user data. The Information field resides only in the frame under the Information frame format. It is usually not found in the Supervisory or Unnumbered frame.

One option of HDLC allows the I field to be used with an Unnumbered Information (UI) frame. This is a very important feature of HDLC because it provides a capability to use the Unnumbered frame to achieve a connectionless-mode operation at the link level. Several subsets of HDLC, such as LLC and LAPD, use the UI frame.

The FCS field is used to check for transmission errors between the two data link stations. The FCS field is created by a cyclic redundancy check. I summarize it here. The transmitting station performs modulo 2 division (based on an established polynomial) on the A, C, and I fields plus leading zeros and appends the remainder as the FCS field. In turn, the receiving station performs a division with the same polynomial on the A, C, I, and FCS fields. If the remainder equals a predetermined value, the chances are quite good that the transmission occurred without any errors. If the comparisons do not match, it indicates a probable transmission error, in which case the receiving station sends a negative acknowledgment, requiring a retransmission of the frame.

The control field

Let's return to a more detailed discussion of the control field (C) because it determines how the HDLC controls the communications process. The control field defines the function of the frame and, therefore, invokes the logic to control the movement of the traffic between the receiving and sending stations. The field can be in one of three formats: information, supervisory, and unnumbered.

The *information* format frame is used to transmit end-user data between the two devices. The Information frame also can acknowledge the receipt of data from a transmitting station and can perform certain other functions such as a poll command.

The *supervisory* format frame performs control functions such as the acknowledgment of frames, the request for the retransmission of frames, and the request for the temporary suspension of the transmission frames. The actual usage of the supervisory frame is dependent on the operational mode of the link (normal response, asynchronous balanced, or asynchronous response mode).

The *unnumbered* format also is used for control purposes. The frame is used to perform link initialization, link disconnection, and other link control functions. The frame uses five bit positions, which allows for up to 32 commands and 32 responses. The particular type of command and response depends on the HDLC class of procedure.

The actual format of the HDLC determines how the control field is coded and used. The simplest format is the Information transfer format. The N(S) (send sequence) number indicates the sequence number associated with a transmitted frame. The N(R) (receive sequence) number indicates the sequence number that is expected at the receiving site.

Piggybacking, flow control, and accounting for traffic

The HDLC maintains accountability of the traffic and controls the flow of frames by the state variables and sequence numbers. The traffic at both the transmitting and receiving sites is controlled by state variables. The transmitting site maintains a send state variable [V(S)], which is the sequence number of the next frame to be transmitted. The receiving site maintains a receive state variable [V(R)], which contains the number that is expected to be in the sequence number of the next frame. The V(S) is incremented with each frame transmitted and placed in the send sequence field in the frame.

Upon receiving the frame, the receiving site checks the send sequence number with its V(R). If the CRC passes and if V(R) equals N(S), it increments V(R) by 1, places the value in the sequence number field in a frame, and sends it to the original transmitting site to complete the accountability for the transmission.

If the V(R) does not match the sending sequence number in the frame (or the CRC does not pass), an error has occurred, and a reject or selective reject with a value in V(R) is sent to the original transmitting site. The V(R) value informs the transmitting DTE of the next frame that it is expected to send (i.e., the number of the frame to be retransmitted).

The poll/final bit

The fifth bit position in the control field is called the P/F, or poll/final, bit. It is recognized only when it is set to 1 and is used by the primary and secondary stations to provide a dialogue with each other:

- The primary station uses the P bit = 1 to solicit a status response from a secondary station. The P bit signifies a poll.

- The secondary station responds to a P bit with data or a status frame and with the F bit = 1. The F bit also can signify end of transmission from the secondary station under Normal Response Mode (NRM).

The P/F bit is called the P bit when used by the primary station and is called the F bit when used by the secondary station. Most versions of HDLC permit one P bit (awaiting an F bit response) to be outstanding at any time on the link. Consequently, a P set to 1 can be used as a checkpoint. That is, the P = 1 means "Respond to me, because I want to know your status." Checkpoints are quite important in all forms of automation. It is the machine's way of clearing up ambiguity and perhaps discarding copies of previously transmitted frames. Under some versions of HDLC, the device might not proceed further until the F bit frame is received, but other versions of HDLC (such as LAPB) do not require the F bit frame to interrupt the full-duplex operations.

How does a station know if a received frame with the fifth bit equal to 1 is an F or P bit? After all, it is in the same bit position in all frames. HDLC provides an elegantly simple solution. The fifth bit is a P bit and the frame is a command if the address field contains the address of the receiving station; it is an F bit and the frame is a response if the address is that of the transmitting station.

This distinction is quite important because a station might react quite differently to the two types of frames. For example, a command (address of receiver, P = 1) usually requires the station to send back specific types of frames.

A summary of the addressing rules follows:

- A station places its own address in the address field when it transmits a response.

- A station places the address of the receiving station in the address field when it transmits a command.

HDLC commands and responses

Table A.1 shows the HDLC commands and responses. They are briefly summarized in the following paragraphs.

The *receive ready* (RR) is used by the primary or secondary station to indicate that it is ready to receive an information frame and/or acknowledge previously received frames by using the N(R) field. The primary station also can use the Receive Ready command to poll a secondary station by setting the P bit to 1.

The *receive not ready* (RNR) frame is used by the station to indicate a busy condition. This informs the transmitting station that the receiving station is unable to accept additional incoming data. The RNR frame can acknowledge previously transmitted frames by using the N(R) field. The busy condition can be cleared by sending the RR frame.

The *selective reject* (SREJ) is used by a station to request the retransmission of a single frame that is identified in the N(R) field. This field also performs inclusive acknowledgment; all information frames numbered up to N(R)–1 are acknowledged. Once the SREJ has been transmitted, subsequent frames are accepted and held for the retransmitted frame. The SREJ condition is cleared upon receipt of an I frame with an N(S) equal to V(R).

An SREJ frame must be transmitted for each erroneous frame; each frame is treated as a separate error. Remember that only one SREJ frame can be outstanding at a time. Therefore, to send a second SREJ would contradict the first SREJ because all of the I frames with N(S) lower than N(R) of the second SREJ would be acknowledged.

TABLE A.1 HDLC Control Field Format

Control field bit format	Encoding								Commands	Responses
	1	2	3	4	5	6	7	8		
Information	0	—	N(S)	—	•	—	N(R)	—	I	I
Supervisory	1	0	0	0	•	—	N(R)	—	RR	RR
	1	0	0	1	•	—	N(R)	—	REJ	REJ
	1	0	1	0	•	—	N(R)	—	RNR	RNR
	1	0	1	1	•	—	N(R)	—	SREJ	SREJ
Unnumbered	1	1	0	0	•	0	0	0	UI	UI
	1	1	0	0	•	0	0	1	SNRM	
	1	1	0	0	•	0	1	0	DISC	RD
	1	1	0	0	•	1	0	0		UP
	1	1	0	0	•	1	1	0		UA
	1	1	0	1	•	0	0	0	NR0	NR0
	1	1	0	1	•	0	0	1	NR1	NR1
	1	1	0	1	•	0	1	0	NR2	NR2
	1	1	0	1	•	0	1	1	NR3	NR3
	1	1	1	0	•	0	0	0	SIM	RIM
	1	1	1	0	•	0	0	1		FRMR
	1	1	1	1	•	0	0	0	SARM	DM
	1	1	1	1	•	0	0	1	RSET	
	1	1	1	1	•	0	1	0	SARME	
	1	1	1	1	•	0	1	1	SNRME	
	1	1	1	1	•	1	0	0	SABM	
	1	1	1	1	•	1	0	1	XID	XID
	1	1	1	1	•	1	1	0	SABME	

LEGEND

I	Information	NR0	Nonreserved 0
RR	Receive Ready	NR1	Nonreserved 1
REJ	Reject	NR2	Nonreserved 2
RNR	Receive Not Ready	NR3	Nonreserved 3
SREJ	Selective Reject	SIM	Set Initialization Mode
UI	Unnumbered Information	RIM	Request Initialization Mode
SNRM	Set Normal Response Mode	FRMR	Frame Reject
DISC	Disconnect	SARM	Set Async. Response Mode
RD	Request Disconnect	SARME	Set ARM Extended Mode
UP	Unnumbered Poll	SNRM	Set Normal Response Mode
RSET	Reset	SNRME	Set NRM Extended Mode
XID	Exchange Identification	SABM	Set Async. Balance Mode
DM	Disconnect Mode	SABME	Set ABM Extended Mode
•	The P/F Bit		

The Reject (REJ) is used to request the retransmission of frames starting with the frame numbered in the N(R) field. Frames numbered N(R)−1 are all acknowledged.

The Unnumbered Information (UI) format allows for retransmission of user data in an unnumbered (i.e., unsequenced) frame. The UI frame actually is a form of connectionless-mode link protocol in that the absence of the N(S) and N(R) fields precludes flow-controlling and acknowledging frames. The IEEE 802.2 logical link control (LLC) protocol uses this approach with its LLC type 1 version of HDLC.

The Request Initialization Mode (RIM) format is a request from a secondary station for initialization to a primary station. Once the secondary station sends RIM, it can monitor frames but can respond to only SIM, DISC, TEST, or XID.

The Set Normal Response Mode (SNRM) places the secondary station in the Normal Response Mode (NRM). The NRM precludes the secondary station from sending any unsolicited frames. This means that the primary station controls all of the frame flow on the line.

The Disconnect (DISC) places the secondary station in the disconnected mode. This command is valuable for switched lines; the command provides a function similar to hanging up a telephone. UA is the expected response.

The Disconnect Mode (DM) is transmitted from a secondary station to indicate it is in the disconnect mode (not operational).

The Test (TEST) frame is used to solicit testing responses from the secondary station. HDLC does not stipulate how the TEST frames are to be used. An implementation can use the I field for diagnostic purposes, for example.

The Set Asynchronous Response Mode (SARM) allows a secondary station to transmit without a poll from the primary station. It places the secondary station in the information transfer state (IS) of ARM.

The Set Asynchronous Balanced Mode (SABM) sets mode to ABM, in which stations are peers with each other. No polls are required to transmit because each station is a combined station.

The Set Normal Response Mode Extended (SNRME) sets SNRM with two octets in the control field. This is used for extended sequencing and permits the N(S) and N(R) to be 7 bits long, thus increasing the window to a range of 1 to 127.

The Set Asynchronous Balanced Mode Extended (SABME) sets SAMB with two octets in the control field for extended sequencing.

The Unnumbered Poll (UP) polls a station without regard to sequencing or acknowledgment. A response is optional if the poll bit is set to 0. This frame provides for one response opportunity.

The Reset (RESET) is used as follows: the transmitting station resets its N(S) and the receiving station resets its N(R). The command is used for recovery. Previously unacknowledged frames remain unacknowledged.

HDLC timers

The vendors vary in how they implement link-level timers in a product. HDLC defines two timers: T1 and T2. Most implementations use T1 in some fashion. T2 is used but not as frequently as T1. The timers are used in the following manner:

- *T1*. A primary station issues a P bit and checks to determine whether a response is received to the P bit within a defined time. This function is controlled by the timer T1 and is called the "wait for F" time-out.
- *T2*. A station in the ARM mode that issues I frames checks to determine whether acknowledgments are received within a timer period. This function is controlled by timer T2 and is called "wait for N(R)" time-out.

Because ARM is not used much today, timer T1 typically is invoked to handle the T2 functions.

HDLC Schema and HDLC "Subsets"

Many other link protocols are derived from HDLC. The practice has proved to be quite beneficial to the industry because it has provided a "baseline" link control standard from which to operate. In some companies, their existing HDLC software has been copied and modified to produce HDLC variations and subsets for special applications. However, be aware that, while these link control systems are referred to as subsets, they sometimes include other capabilities not found in HDLC.

The major published subsets of HDLC are summarized in this section. The overall HDLC schema is shown in Figure A.2. Two options are provided for unbalanced links normal response mode (UN) and asynchronous response mode (UA) and one for balanced asynchronous balanced mode (BA).

To classify a protocol conveniently, the terms UN, UA, and BA are used to denote which subset of HDLC is used. In addition, most subsets use the functional extensions. For example, a protocol classified as UN 3,7 uses the unbalanced Normal Response Mode option and the selective reject and extended address functional extensions.

From an examination of Figure A.2, it is evident that HDLC provides a wide variety of options. Consequently, the full range of functions are not implemented as a single product. Rather, a vendor chooses the subset that best meets the need for the link protocol. You should be aware that using an HDLC product does not guarantee link compatibility with another vendor's HDLC product because each might implement a different subset of HDLC. Furthermore, some vendors implement features not found in the HDLC standards. When in doubt, read the manuals.

Link Access Procedure (LAP)

LAP is an earlier subset of HDLC. LAP is based on the HDLC Set Asynchronous Response Mode (SARM) command on an unbalanced configuration. It

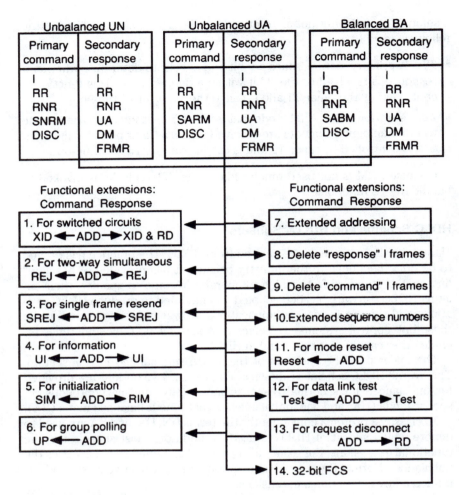

Figure A.2 The HDLC schema.

is classified UA 2,8 except that it does not use the DM response. LAP still is used to support some X.25 network links.

To establish an LAP data link, the sending end (primary function) transmits an SARM in the control field to the receiving end (secondary function). Concurrent with the transmission of the SARM, the primary function will start a no-response timer (T1). When the secondary function receives the SARM correctly, it transmits an acknowledgment response (UA: Unnumbered Acknowledgment). Receipt of the UA by the primary function confirms the initiation of one direction of the link and resets the T1 timer. The receipt of the SARM in a given direction will be interpreted by the secondary function as a request to initiate the other direction of transmission so the procedure may be repeated in the other direction at the discretion of the secondary function.

Link Access Procedure, Balanced (LAPB)

LAPB is used by many private and public computer networks throughout the world. LAPB is classified as a BA 2,8 or BA 2,8,10 subset of HDLC. Option 2 provides for simultaneous rejection of frames in a two-way transmission mode.

Option 8 does not permit the transmitting of information in the response frames. This restriction presents no problem in an asynchronous balanced mode because the information can be transferred in command frames, and because both stations are combined stations, both can transmit commands. Moreover, with LAPB, the sending of a command frame with the P bit = 1 occurs when the station wants a "status" frame and not an information frame. Consequently, the responding station is not expected to return an I field.

LAPB is the link protocol layer for an X.25 network. It is used extensively worldwide and is found in many vendor's ports on a chip with the X.25 network level software. An extensive discussion on LAPB can be found in my book *The X.25 Protocol*, which was published by the IEEE Computer Society.

Logical Link Control (LLC)

LLC (IEEE 802.2 and ISO 8802) is a standard sponsored by the IEEE 802 standards committee for local area networks (LANs). The standard permits the interfacing of a LAN to other local networks as well as to a wide area network. LLC uses a subclass of the HDLC superset. LLC is classified as BA-2,4.

LLC permits three types of implementations of HDLC: type 1, using the UI frame (Unacknowledged Connectionless Service); type 2, using the conventional I frame (Acknowledged Connection-Oriented Service); and type 3, using AC frames (Acknowledged Connectionless Service). See Table A.2.

LLC is intended to operate over a peer-to-peer multipoint channel using the UI or SABME frames. Therefore, each frame contains the address of both the sending and receiving station.

Link Access Procedure, D channel (LAPD)

LAPD is another subset of the HDLC structure, although it has extensions beyond HDLC. It is derived from LAPB. LAPD is used as a data link control for the integrated services digital network (ISDN).

ISDN provides LAPD to allow DTEs to communicate with each other across the ISDN D Channel. (Many users want LAPD for B channels as

TABLE A.2 The HDLC Implementation Types

		Commands		Responses
Type 1		UI	—	
		XID	—	XID
		TEST	—	TEST
Type 2	(I format)	I	—	I
	(S format)	RR	—	RR
		RNR	—	RNR
		REJ	—	REJ
	(U format)	SABME	—	
		DISC	—	UA
			—	DM
			—	FRMR
Type 3		AC		AC

well.) It is specifically designed for the link across the ISDN user-network interface.

LAPD has a very similar frame format to HDLC and LAPB. Moreover, it provides for unnumbered, supervisory, and information transfer frames. LAPD also allows a Modulo 128 operation. The control octet to distinguish between the Information, Supervisory, and Unnumbered formats is identical to HDLC.

LAPD provides for two octets for the address field. This is valuable for multiplexing multiple functions onto the D channel. Each ISDN basic access can support up to eight stations. The address field is used to identify the specific terminal and service access point (SAP). The address field contains the address field extension bits, a command/response (C/R) indication bit, a service access point identifier (SAPI), and a terminal end-point identifier (TEI). These entities are discussed in the following paragraphs.

The purpose of the address field extension is to provide more bits for an address. The presence of a 1 in the first bit of an address field octet signals that it is the final octet of the address field. Consequently, a two-octet address would have a field address extension value of 0 in the first octet and a 1 in the second octet. The address field extension bit allows the use of both the SAPI in the first octet and the TEI in the second octet.

The C/R field bit identifies the frame as either a command or a response. The user side sends commands with the C/R bit set to 0. It responds with the C/R bit set to 1. The network does the opposite—it sends commands with C/R set to 1 and responses with C/R set to 0.

The service access point identifier (SAPI) identifies the point where the data link layer services are provided to the layer above (that is, layer 3). (The concept of the SAPI is covered in chapter 1.)

The terminal end-point identifier (TEI) identifies a specific connection within the SAP. It can identify either a single terminal (TE) or multiple terminals. The TEI is assigned by a separate assignment procedure. Collectively, the TEI and SAPI are called the data link connection identifier (DLCI), which identifies each data link connection on the D channel. As stated earlier, the control field identifies the type of frame as well as the sequence numbers used to maintain windows and acknowledgments between the sending and receiving devices. Presently, the SAPI values and TEI values are allocated as shown in Table A.3.

Two commands and responses in LAPB do not exist in the HDLC schema. These are sequenced information 0 (SI0) and sequenced information 1 (SI1). The purpose of the SI0 and SI1 commands is to transfer information using sequentially acknowledged frames. These frames contain information fields that are provided by layer 3. The information commands are verified by the means of the end (SI) field. The P bit is set to 1 for all SI0 and SI1 commands. The SI0 and SI1 responses are used during single frame operation to acknowledge the receipt of SI0 and SI1 command frames and to report the loss of frames or any synchronization problems. LAP does not allow information fields to be placed in the SI0 and SI1 response frames. Obviously, information fields are in the SI0 and SI1 command frames.

LAPD differs from LAPB in a number of ways. The most fundamental difference is that LAPB is intended for point-to-point operating [user DTE-to-packet exchange (DCE)]. LAPD is designed for multiple access on the link. The other major differences are summarized as follows:

- LAPB and LAPD use different timers.

- As explained earlier, the addressing structure differs.

- LAPD implements the HDLC unnumbered information frame (UI).

- LAPB uses only the sequenced information frames.

TABLE A.3 The Allocation of SAPI and TEI Values

SAPI	
Value	Related entity
0	Call control procedures
16	Packet procedures
32–47	Reserved for national use
63	Management procedures
Others	Reserved

TEI	
Value	User type
0–63	Nonautomatic assignment
64–126	Automatic assignment

TABLE A.4 LAPD Primitives

Primitives	Function
DL-ESTABLISH (level 2/3 boundary)	Issued on the establishment of frame operations
DL-RELEASE (level 2/3 boundary)	Issued on the termination of frame operations
DL-DATA (boundary)	Used to pass data between layers (level 2/3 with acknowledgments)
DL-UNIT-DATA (level 2/3 boundary)	Used to pass data with no acknowledgments
MDL-ASSIGN (level management/2 boundary)	Used to associate TEI value with a specified end point
MDL-REMOVE (level management/2 boundary)	Removes the MDL-ASSIGN
MDL-ERROR (level management/2 boundary)	Associated with an error that cannot be corrected by LAPD
MDL-UNIT-DATA (level management/2 boundary)	Used to pass data with no acknowledgments
PH-DATA (level 2/1 boundary)	Used to pass frames across layers
DH-ACTIVATE (level 2/1 boundary)	Used to set up physical link
PH-DEACTIVATE (level 2/1 boundary)	Used to deactivate physical link

LAPD primitives. LAPD uses a number of primitives for its communications with the network layer, the physical layer, and a management entity which resides outside both layers. The primitives are summarized in Table A.4.

LAPM with V.42

This section examines a new ITU-T protocol—LAPM. V.42 implements a link control protocol called LAPM. It is based on the HDLC family of protocols and also was written from the LAPB specification, which is part of X.25.

The principal difference between LAPM and a conventional HDLC implementation relates to the use of the address field. The address field consists of the data link identifier, the C/R bit, and the address extension bit. The C/R bit is a command/response bit that identifies the frame as either a command or response. The DLCI value is used to transfer information between the X.24 interfaces. Currently, DLCI is set to 0 to identify a DTE-DTE interface. Other values are permitted within the limits defined in the recommendation. The EA bit can be set to 1 to designate another octet for DLCI.

B

Tutorial on ISDN

You should remember that ITU-T intends the ISDN to complement its V Series Recommendations for data services. The complementary functions include user facilities, quality of service features, and call progress signals such as those established in some of the X Series Recommendations (X.2 and X.96).

Although less explicitly stated, it is evident in several parts of this book that ITU-T intends to make the transition from analog to digital services as transparent as possible to the end user. For example, the terminal adapter (TA) is an integral part of many of the V and X series recommendations. Its function is to provide a transition interface from the current analog interfaces to the integrated services digital network (ISDN).

The ISDN Terminal

To begin the analysis of ISDN, consider the end-user ISDN terminal in Figure B.1. This device (called data terminal equipment, or DTE, in this book) is identified by the ISDN term TE1 (terminal equipment, Type 1). The TE1 connects to the ISDN through a twisted-pair four-wire digital link. This link uses time division multiplexing (TDM) to provide three channels, designated as the B, B, and D channels (or 2B+D). The B channels operate at a speed of 64 kbit/s; the D channel operates at 16 kbit/s. The 2B+D is designated as the basic rate interface. ISDN also allows up to eight TE1s to share one 2B+D channel.

Figure B.2 illustrates the format for the basic access D channel. The information (I) field carries upper-layer information. With an ISDN connection, it carries the ISDN network layer. It also could carry an X.25 packet.

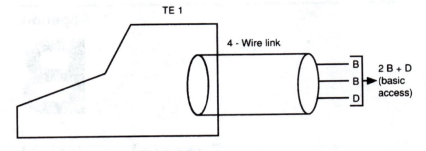

B channels = 64 Kbit/s each
D channel = 16 Kbit/s

2B + D = 144 Kbit/s as the basic rate

TE1 = terminal equipment 1 (an ISDN device)

Figure B.1 ISDN basic access.

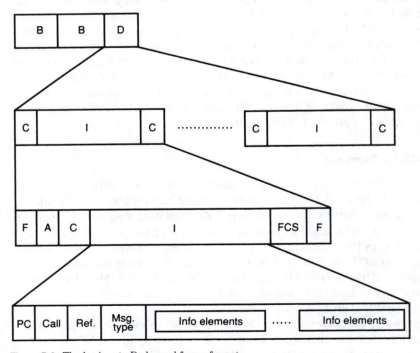

Figure B.2 The basic rate D channel frame format.

Figure B.3 illustrates other ISDN options. In this scenario, the user DTE is a TE2 device, which currently is in use in such equipment as IBM 3270 terminals, telex devices, etc. The TE2 connects to the terminal adapter (TA), which allows non-ISDN terminals to operate over ISDN lines. The user side of the TA typically uses a conventional physical-layer interface, such as EIA-232, or the V-series specification that was discussed earlier in this book. It is packaged like an external modem or as a board that plugs into an expansion slot on the TE2 devices. The EIA or V Series interface is called the R interface in ISDN terminology.

Basic Access and Primary Access

The TA and TE2 devices are connected through the basic access to either an ISDN NT1 or NT2 device (NT is network termination). The NT1 is a customer premise device that connects the four-wire subscriber wiring to the conventional two-wire local loop. ISDN allows up to eight terminal devices to be addressed by NT1.

The NT1 is responsible for the physical-layer functions (of OSI), such as signalling synchronization and timing. It provides a user with a standardized interface.

The NT2 is a more intelligent piece of customer premise equipment. It typically is found in a digital PBX and contains the layer 2 and 3 protocol

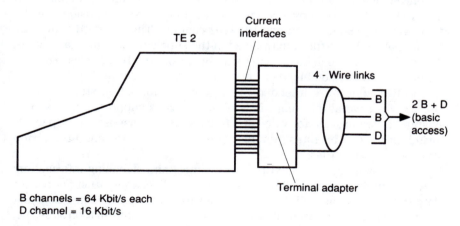

B channels = 64 Kbit/s each
D channel = 16 Kbit/s

Terminal adapter (TA) = protocol converter

TE2 = terminal equipment 2 (current equipment)

Figure B.3 The ISDN terminal adapter.

functions. The NT2 device is capable of performing concentration services. It multiplexes 23 B+D channels onto the line at a combined rate of 1.544 Mbit/s. This function is called the ISDN primary rate access.

The NT1 and NT2 devices can be combined into a single device called NT12. This device handles the physical, data link, and network layer functions.

In summary, the TE equipment is responsible for user communications and the NT equipment is responsible for network communications.

ISDN Reference Points and Interfaces

The reference points are logical interfaces between the functional groupings. The S reference point is the 2B+D interface into the NT1 or NT2 device. The T interface is the reference point on the customer side of the NT1 device. It is the ISDN "plug in the wall." It is the same as the S interface on the basic rate access lines. The U interface is the reference point for the two-wire side of the NT1 equipment. It separates an NT1 from the line termination (LT) equipment. The V reference point separates the line termination (LT) from the exchange termination (ET) equipment.

ISDN Channels

The most common ISDN interface supports a bit rate of 144 kbit/s. The rate includes two 64-kbit/s B channels, and one 16-kbit/s D channel. In addition to these channels, ISDN provides for framing control and other overhead bits, which totals to a 192-kbit/s bit rate. The 144-kbit/s interfaces operate synchronously in the full-duplex mode over the same physical connector. The 144-kbit/s signal provides time-division multiplexed provisions for the two 64-kbit/s channels and one 16-kbit/s channel. The standard allows the B channels to be further multiplexed in the subchannels. For example, 8-, 16-, or 32-kbit/s subchannels can be derived from the B channels. The two B channels can be combined or broken down as the user desires.

The B channels are intended to carry user information streams. They provide for several different kinds of applications support. For example, channel B can provide for voice at 64 kbit/s, data transmission for packet-switch utilities at bit rates less than or equal to 64 kbit/s, and broad-band voice at 64 kbit/s or less.

The D channel is intended to carry control and signalling information; however, in certain cases, ISDN allows for the D channel to support user data transmission as well. However, be aware that the B channel does not carry signalling information. ISDN describes signalling information as s-type, packet data as p-type, and telemetry as t-type. The D channel can carry all of these types of information through statistical multiplexing.

ISDN Layers

The ISDN approach is to provide an end user with full support through the seven layers of the OSI Model. In so doing, ISDN is divided into two kinds of services: the bearer services, which are responsible for providing support to the lower three levels of the seven-layer standard, and teleservices (for example, telephone, Teletex, Videotex message handling), which are responsible for providing support through all seven layers of the model and generally making use of the underlying lower-level capabilities of bearer services. The services are referred to as low- and high-layer functions, respectively. The ISDN functions are allocated according to the layering principles of the OSI and ITU-T standards.

I hope that this brief tutorial on the major ISDN terms and concepts will be sufficient for you to cope with these ideas as they relate to the V Series Recommendations.

C

ITU-T V Series Recommendations

This appendix lists all the ITU-T V Series Recommendations. These documents can be obtained at:

ITU-T Sales Department
Place des Nations
CH-1211 Geneve 20 Suisse
Telephone: 730.52.85
Fax: 730.51.94

Other organizations (commercial services) offer these documents, but the ITU-T prices are lower than the prices of these organizations. Some of the documents are available in some electronic libraries, but be aware that the material is copyrighted.

General: V.1 through V.8

V.1: Equivalence between binary notation symbols and the significant conditions of a two-condition code.

V.2: Power levels for data transmission over telephone lines.

V.4: General structure of international alphabet No. 5 (IA5) code for character-oriented data transmission over public telephone networks.

V.5: Standardization of data signalling rates for synchronous data transmission in the general switched telephone network (NLIF).

V.6: Standardization of data signalling rates for synchronous data transmission on leased telephone-type circuits (NLIF).

V.7: Definitions of terms concerning data communications over the telephone network.

V.8: Procedures for starting sessions of data transmission over the switched telephone network (UCFA).

Interfaces and Voice-Band Modems: V.10 through V.34

V.11: Electrical characteristics for balanced double-current interchange circuits operating at data signalling rates up to 10 Mbit/s.

V.13: Simulated carrier control.

V.14: Transmission of start-stop characters over synchronous bearer channels.

V.15: Use of acoustic coupling for data transmission.

V.16: Medical analogue data transmission modems.

V.17: A two-wire modem for facsimile applications with rates up to 14,400 bits/s.

V.18: Operational and interworking requirements for modems operating in the text telephone mode.

V.19: Modems for parallel data transmission using telephone signalling frequencies.

V.20: Parallel data transmission modems standardized for universal use in the general switched telephone network.

V.21: 300 bit/s duplex modem standardized for use in the general switched telephone network (GSTN).

V.22: 1200 bit/s duplex modem standardized for use in the general switched telephone network and on point-to-point, two-wire leased telephone-type circuits.

V.22 bis: 2400 bit/s duplex modem using the frequency division technique standardized for use on the general switched telephone network and on point-to-point, two-wire leased telephone-type circuits.

V.23: 600/1200 baud modem standardized for use in the general switched telephone network.

V.24: List of definitions for interchange circuits between data terminal equipment and data circuit-terminating equipment.

V.25: Automatic answering equipment and/or parallel automatic calling equipment on the general switched telephone network, including procedures for disabling of echo control devices for both manually and automatically established calls.

V.25 bis: Automatic calling and/or answering equipment on the general switched telephone network (GSTN) using the 100-series interchange circuits.

V.26: 2400 bit/s modem standardized for use on four-wire leased telephone-type circuits.

V.26 bis: 2400/1200 bit/s modem standardized for use in the general switched telephone network.

V.26 ter: 2400 bit/s duplex modem using the echo cancellation technique standardized for use on the GSTN and on point-to-point, two-wire leased telephone-type circuits.

V.27: 4800 bit/s modem with manual equalizer standardized for use on leased telephone-type circuits.

V.27 bis: 4800/2400 bit/s modem with automatic equalizer standardized for use on leased telephone-type circuits.

V.27 ter: 4800/2400 bit/s modem standardized for use in the general switched telephone network.

V.28: Electrical characteristics for unbalanced double-current interchange circuits.

V.29: 9600 bit/s modem standardized for use on point-to-point, four-wire leased telephone-type circuits.

V.31: Electrical characteristics for single-current interchange circuits controlled by contact closure.

V.31 bis: Electrical characteristics for single-current interchange circuits using optocouplers.

V.32: A family of 2-wire, duplex modems operating at data signalling rates of up to 9600 bit/s for use on the GSTN and on leased telephone-type circuits.

V.32 bis: A duplex modem operating at data signalling rates of up to 14,400 bit/s for use on the general switched telephone network and on leased point-to-point, two-wire telephone-type circuits.

V.33: 14,400 bit/s modem standardized for use on point-to-point four-wire leased telephone-type circuits.

V.34: A modem operating at data signalling rates of up to 28,800 bit/s for use on the general switched telephone network and on leased point-to-point 2-wire telephone-type circuits.

Wide-Band Modems: V.35 through V.38

V.35: Data transmission at 48 kbit/s using 60- to 108-kHz group band circuits (NLIF).

V.36: Modems for synchronous data transmission using 60- to 108-kHz group band circuits.

V.37: Synchronous data transmission at a data signalling rate higher than 72 kbit/s using 60- to 108-kHz group band circuits.

V.38: A 48/56/64 kbit/s data circuit terminating equipment standardized for use on digital point-to-point leased circuits.

Error Control: V.40 through V.42

V.40: Error indication with electromechanical equipment (NLIF).

V.41: Code-independent error-control system.

V.42: Error-correcting procedures for DCEs using synchronous-to-asynchronous conversion.

V.42 bis: Data compression procedures for DCEs using error correcting procedures.

Transmission Quality and Maintenance: V.50 through V.58

V.50: Standard limits for transmission quality of data transmission.

V.51: Organization of the maintenance of international telephone-type circuits used for data transmission.

V.52: Characteristics of distortion and error-rate measuring apparatus for data transmission.

V.53: Limits for the maintenance of telephone-type circuits used for data transmission.

V.54: Loop test devices for modems.

V.55: Specification for an impulsive noise-measuring instrument for telephone-type circuits.

V.56: Comparative tests of modems for use over telephone-type circuits.

V.57: Comprehensive data test set for high data signalling rates.

V.58: Management information model for V series DCEs.

Interworking with Other Networks: V.100 through V.230

V.100: Interconnection between public data networks (PDNs) and the public switched telephone network (PSTN).

V.110: Support of data terminal equipment (DTEs) with V-Series type interfaces by an integrated services digital network (ISDN).

V.120: Support by an ISDN of data terminal equipments (DTEs) with V-Series type interfaces with provision for statistical multiplexing.

V.230: General data communications interface layer 1 specification.

Epilogue

This book has provided a tutorial summary and reference guide to the ITU-T V Series Recommendations. I hope that you have gained an understanding of these important standards and how they can be used in data communications systems. As stated several times, this book should not serve as a substitute for the ITU-T documents. However, you now should be able to read the technical specification with more ease.

It is stated by some people in the industry that, in the not too distant future, products such as modems, multiplexers, and data service units will no longer implement V Series functions. That might be so; however, experience tells us that the movement away from a technology as capital intensive as the V Series might take many years. I believe that the use of the V Series will remain in the industry for a considerable period, if for no other reason than that it is proving to be expensive to move out of the analog technology. Secondly, the ITU-T has done an excellent job of providing transition protocols through the interworking specifications. They allow a user to keep the current V Series-based equipment, yet interface into the newer ISDN-based machines.

It is equally clear that the continued use of duplicate technologies (analog and digital) makes no sense from the standpoints of costs and performance. Few people argue with the statement that digital systems are replacing analog systems at a fairly rapid rate. Eventually this book probably will find itself on the history shelves.

However, for the near future, the V Series will continue to play a prominent role in an organization's data communications systems and networks. The ITU-T has done a laudable job in fostering and directing the development of these standards. Their use has led to decreased costs of communication interfaces, as well as increased performance of data communications systems and networks.

Index

A

abandon call, 71
acoustic couplers, 59
amplitude modulation (AM), 41-42
  quadrature, 44, 95
analog signals, 29-30
answer-detection pattern (ADP), 125
application layer, 9
asynchronous balanced mode (ABM), 171
asynchronous response mode (ARM), 171
asynchronous transmission, 27-28
automatic call and answer, 74
automatic dial and answer, 73

B

backup switching, 66
backward channel ready, 67
backward channel received line signal de-
    tector, 67
backward channel signal quality detector, 67
bandwidth, 32-33, 45
baseband signals, 33
baud rate, 39
bit rate, 39
bit stuffing, 173
broadband signals, 33

C

cable
  balanced twisted pair, 20
  connectors and, 45-46
  shielded twisted pair, 20-21
  telephone twisted pair, 20

twisted pair, 19
  unbalanced twisted pair, 20
  unshielded twisted pair, 20-21
call request, 71
calling indicator, 67
carrier frequency, 18
cellular modem, 47
clocking, 28-29
common return, 61, 64, 71
communications
  establishing between DCEs, 124
  line characteristics, 21-24
  media, 19-21
compression schemes, vendor-specific, 139-
    140
connect data set to line, 65
connectors, 45-46
constellation pattern, 44
control function, 122
cycles, 31-32

D

data channel received line signal detector, 65
data circuit-terminating equipment (DCE),
    2-3, 5, 17, 53
  establishing communications between, 124
data communications, 7
  synchronizing components, 25-27
data line occupied, 71
data link connection identifier (DLCI), 183
data link control identifier (DLCI), 123
data link layer, 8-9
data set ready, 65
data signal quality detector, 65
data signalling rate selector, 66

ABOUT THE AUTHOR

Uyless Black is the founder of the Information Engineering Institute in Virginia. He is the author of numerous books and articles on computer communications and lectures and consults worldwide on the subject. He is McGraw-Hill's series advisor for the *Uyless Black Series on Computer Communications*.